Deploying Optical Networking Components

Gilbert Held

McGraw-Hill
New York · Chicago · San Francisco · Lisbon
London · Madrid · Mexico City · Milan · New Delhi
San Juan · Seoul · Singapore
Sydney · Toronto

Cataloging-in-Publication Data is on file with the Library of Congress

McGraw-Hill

A Division of The ***McGraw-Hill*** *Companies*

1 2 3 4 5 6 7 8 9 0 DOC/DOC 0 9 8 7 6 5 4 3 2 1

ISBN 0-07-137505-8

The sponsoring editor for this book was Marjorie Spencer, the editing supervisor was Steven Melvin, and the production supervisor was Sherri Souffrance. It was set in Vendome ICG by Paul Scozzari and Charles Nappa of McGraw-Hill's Professional Book Group composition unit, Hightstown, N.J.

Printed and bound by R. R. Donnelley & Sons Company.

McGraw-Hill books are available at special quantity discounts to use as premiums and sales promotions, or for use in corporate training programs. For more information, please write to the Director of Special Sales, McGraw-Hill, 2 Penn Plaza, New York, NY 10121-2298. Or contact your local bookstore.

This book is printed on recycled, acid-free paper containing a minimum of 50% recycled, de-inked fiber.

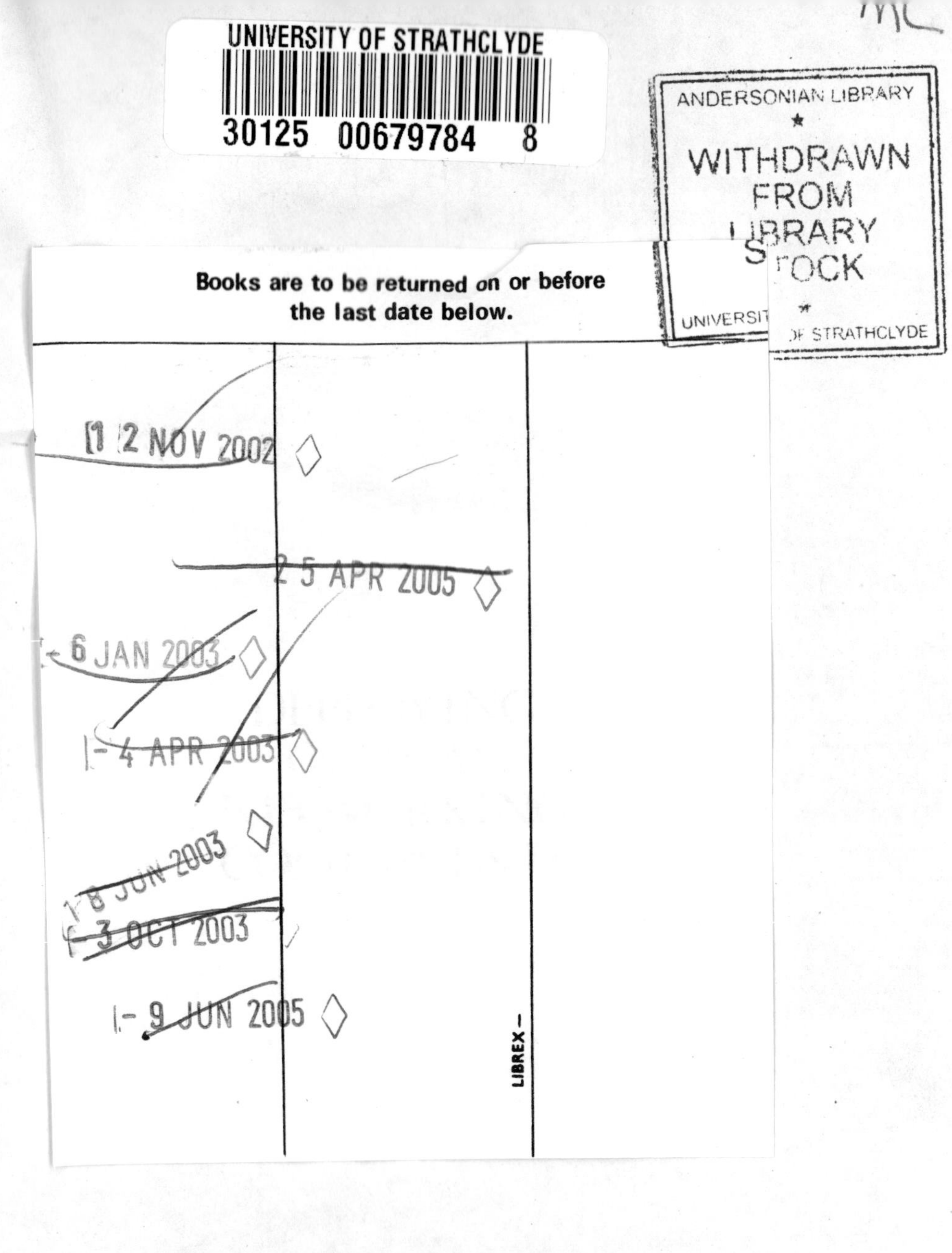
Books are to be returned on or before
the last date below.
12 NOV 2002
25 APR 2005
6 JAN 2003
4 APR 2003
18 JUN 2003
3 OCT 2003
9 JUN 2005
LIBREX

CONTENTS

PREFACE

Today we are witnessing a revolution in the field of communications. That revolution involves the use of light as a mechanism to convey information along an optical fiber. Although the use of light along an optical fiber dates to experimental efforts during the 1930s and was successfully implemented during the 1970s, it wasn't until the turn of the millennium that the use of optical networking exploded.

As we entered the new millennium, optical networking represented the fastest-growing segment in the rapidly developing field of communications. Because of their high bandwidth and low noise, optical fibers are well suited to transport information. While many articles in trade publications have focused on the use of optical networking to provide the infrastructure necessary to cope with the increasing use of the Internet, as a famous announcer would say, "That's only part of the story." The real story associated with optical networking is its widespread use in local and wide area networks (LANs and WANs) and, more recently, in intrabuilding communications.

The use of optical networking devices is beginning to approach ubiquity within the office environment and is expected to extend to the home within a few years. In the office environment it is quite common to encounter the use of optical repeaters to transfer data at distances greater than those possible via copper cable. In addition to repeaters, the use of optical modems and multiplexers, as well as storage area networks based on the transfer of data over optical fiber, is also becoming common. In the home environment both cable and telephone companies are using fiber-to-the-neighborhood (FTTN), fiber-to-the-curb (FTTC), and fiber-to-the-home (FTTH) systems as mechanisms to provide broadband communications to the residential market. Because of the explosion in the use of optical networking devices, it is important for communications professionals to understand how optical networking occurs as well as the numerous options available. That is the focus of this book.

In this book we will focus on the various components of an optical transmission system and how those components are used to support both effective and efficient transmission via light. Because it is reasonable to expect readers to have both a diverse background and more than likely different networking requirements, this book was written to provide insight into many areas of optical networking. Thus, in this book we will become acquainted with the operation of different optical networking components as well as their use in development of transmission systems for use in WANs, LANs, and other application environments.

As a professional author with over 30 years of hands-on networking experience, I highly value reader feedback. Please feel free to contact me via email at *gil_held@yahoo.com* concerning any comments you wish to share concerning this book. Let me know if I omitted an area or topic that should be considered for the next edition of this book, or if I put too much or too little emphasis on a topic. With that said, grab a Coke, Pepsi, or your cup of tea and follow me as we explore the wonderful world of optical networking.

Gilbert Held
Macon, Georgia

ACKNOWLEDGMENTS

While an author's name appears prominently on most books, the author is just one person in a team responsible for the preparation, production, and marketing effort that results in a title appearing on a bookstore shelf or being advertised in a hardcopy or electronic publication. Thus, I would be remiss if I did not acknowledge the fine efforts of many individuals who are collectively responsible for this book.

As an old-fashioned author who spends a considerable amount of time on airplanes and in hotel rooms, I recognized some time ago that there are limits concerning the use of technology. In particular, air turbulence at 30,000 ft makes it difficult to type or attempt to draft an illustration on a notebook computer, while few hotel rooms have connector receptacles on their electrical outlets that allow the connector plugs to mate in the universal connector traveling kit I use. Recognizing these problems, I turned to the use of the universal word processor: a pencil and paper. While my writing may appear choppy during an encounter with air turbulence at 30,000 ft, I have never lost power to my pen or pencil and do not have to worry about whether my connector plugs will fit into the electrical socket in a hotel to recharge my writing implements. Of course, I now have to worry about finding someone who can interpret my handwriting and convert my writing and drafted illustrations into a professional manuscript. Fortunately, I am able to depend on the skills of Mrs. Linda Hayes. Linda has done wonders interpreting my handwriting and turning my drawings into a professional manuscript, for which I am truly grateful.

While the creation of a professional manuscript is an important part of the book publishing process, I would again be remiss if I did not thank my acquisition editor and the people in the production department for their efforts. Although the term "book acquisition" may appear to be simple, in actuality the process is quite involved and requires the acquisition editor to take the author's proposal and adjust it to the internal organizational format, present it at a formal meeting, and defend its viability. This obviously requires a considerable amount of effort, and once again I am indebted to Marjorie Spencer for shepherding my proposal into a contract.

As a manuscript moves into the production process, it passes through the hands of copyeditors, illustrators, cover and backcopy designers, and other specialists whom I would like to collectively thank. Last but certainly not least, I would like to thank my wife, Beverly, for her patience and understanding during the long nights and lost weekends while I worked researching and writing this book.

CHAPTER 1

Introduction

Imagine working in an environment where you can dynamically obtain extra bandwidth to satisfy different operational requirements. Suppose it is 8:00 A.M. on a Tuesday after a long weekend. As workers arrive and discuss their weekend activities, your organization's requirement for bandwidth may be minimal. As the week progresses, the activity of your organization increases. Customers call in seeking support, employees access a variety of remote databases, and your organization's bandwidth requirements increase.

The current bandwidth requirements of organizations are satisfied through the use of a fixed capacity local loop line linking organizational locations to each other or to different networks. Even when capacity to a degree appears variable, such as when using Frame Relay where it is possible to burst above your committed information rate (CIR), you can burst only up to the fixed line operating rate. Now imagine the use of communications lines connecting each organizational location to a network where each local loop line has such a large—in fact, virtually unlimited—transmission capacity. As your organizational requirements grow and you need more bandwidth, the transmission capacity of the line terminating device is so high that the requirements for additional bandwidth can be satisfied well into the current millennium without requiring a line upgrade. Does this sound like a fantasy? The answer to this question depends on our time reference. While dynamic bandwidth on demand may not be ready for prime time while you read this book, within a few years it should be a reality. The key to its pending availability as well as the ability to obtain practical videoconferencing, multimegabit file transfers in a fraction of a second, movies on demand, and dynamic connections to distance learning centers, and satisfy other bandwidth-intensive applications will be obtained through the deployment of optical networking devices.

Optical Networking and Basic Terminology

Because this book is about optical networking, before we go too far into this topic it might be a good idea to compare and contrast this method of communications with conventional copper-based communications to understand their similarities and differences. In a conventional copper-based communications system a modem or digital service unit (DSU) functions as both the transmitter and the receiver, while the copper conductor represents the transmission medium. In an optical networking environment we still require the three pillars of any type of communications system: transmitter, receiver, and transmission medium. A laser or a light-emitting diode (LED) represents the transmitter, while a light detector, referred to as a *photodetector* or an *optical detector,* represents the receiver. Both are normally solid-state semiconductor devices. As you might expect, the copper cable is replaced by an optical fiber as the transmission medium.

Terminology

In this book we will encounter both new and old terminology as well as some possibly confusing terminology that this section should help you understand. Concerning the latter, throughout this book this author uses the terms *optical fiber* and *fiber optic* interchangeably even though a purist might disagree. Thus, we will consider both glass and plastic to represent optical fibers as well as possible material used in a fiber optic. However, because plastic fiber provides only a fraction of the capability of glass fiber, we will specifically reference the former in this book. Otherwise, all references to fiber will be to glass-based fiber.

Concerning other terminology, while we will review applicable terminology throughout this book, a reminder of prefix values might be in order, especially for a book that covers frequency, bandwidth, and wavelengths that can considerably vary from the normal numbers that we work with on a daily basis. Thus, as a refresher, the prefix *milli* represents 10^{-3}, *micro* represents 10^{-6}, and *nano* represents 10^{-9}, while *kilo* represents 10^{3}, *mega* represents 10^{6}, and *giga* represents 10^{9}. Other prefixes worth noting with respect to the field of optical networking but rarely encountered in everyday life are *pico* (10^{-12}), *femto* (10^{-15}), *tera,* a prefix that represents a trillion (10^{12}), and *peta* (10^{15}), which represents a quadrillion.

Deployment

The deployment of optical networking devices is occurring at a rapid rate within the telephone company and cable television backbone infrastructure into the local neighborhood and onto local area networks (LANs). If you simply count the number of spools of different types of optical fiber being shipped by cable manufacturers, you will easily understand why their stockmarket values appeared during the first half of the year 2000 to represent the liftoff of a Saturn-bound rocket. Similarly, the manufacturers of a variety of optical components such as lasers, light-emitting diodes (LEDs), couplers, optical modems, and multiplexers have been recognized as growth stocks as numerous organizations either acquire optical systems on a turnkey basis or purchase individual components and integrate those components into an optical system to satisfy their organizational requirements.

One common measurement of optical deployment is given in terms of a communications carrier's installed or planned installation of fiber miles. Here the term *fiber miles* represents the length of the fiber conduit installed or to be installed by a network operator multiplied by the illuminated (lit) and dark fiber-optic strands in the conduit. The term *lit* refers to those fibers that transport information, while the term *dark* represents those strands not yet in use. Because a major cost associated with the construction of a fiber-optic network involves acquiring rights-of-way, digging a trench, and installing a conduit, adding dark fiber for later illumination is a common practice. Table 1.1 summarizes the network buildout of six communications carriers in terms of fiber miles.

Although fiber miles represents an important indication of the scope of geographic coverage of a network, it does not indicate the true capacity of a network. Concerning network capacity, it is important to note the type of fiber installed, as certain types of fiber are more suitable than other types for supporting high-speed transmission at data rates of 10 Gbits/s (gigabits per second) and an emerging transmission rate of 40 Gbits/s per strand of optical fiber, topics that will be covered in this book.

Book Focus

As readers have a diverse background and their organizations have different networking requirements, the goal of this book ultimately was to provide a practical guide to the operation and utilization of optical networking devices. In this book we will focus on the various components that make up an optical transmission system, how each component operates, and how all these components are integrated to form an optical net-

TABLE 1.1
Network Buildouts in Terms of Fiber Miles

Network operator	Fiber miles	Network completion date
Aerie Networks	8,885,376	2004 (projected)
AT&T	5,148,000	2001 (projected)
Level 3 Communications	2,304,000	2001 (projected)
Qwest	1,836,480	Completed
Broadwing	1,507,200	Completed
Global Crossing	480,000	Completed

work. As we progress through this book, we will obtain an appreciation for the composition of light and how it flows down different types of optical fiber. We will note the use of different types of light sources that function as a transmitter in an optical system as well as the use of different types of photodetectors that function as light receivers. Through our coverage of operational information, we will obtain a foundation of knowledge that we will use to understand how optical transmission can be used in local and wide area networking environments as well as within a building to satisfy a variety of different networking requirements.

Chapter Focus

Until the late 1990s or so, most persons associated optical networking with wide area networks (WANs) as communications carriers extended their infrastructure to support the growth in the use of the Internet, videoconferencing, and other bandwidth-intensive applications. What was not common knowledge is the fact that optical networking devices can and are being used in the office environment and are gaining momentum in providing communications to residences. To understand the reason for the growth in the use of optical networking devices, the first section of this chapter provides an overview of the rationale behind the use of optical fiber. To ensure that readers do not have a biased impression that everything is rosy and that we should migrate all communications to a light-based system, we will also describe and discuss certain reasons why copper can remain a better choice than the use of optical fiber. Once we have an appreciation for the advantages and disadvantages associated with the use of optical fiber, we will discuss a wide range of applications dependent on the use of light. We will conclude this introductory chapter with a preview of succeeding chapters. You can use this chapter preview by itself or in conjunction with the index to locate specific information of interest.

While the chapters in this book were structured in a sequence to provide an optimum benefit for readers, this author also recognizes the reality of a modern work environment where many persons receive assignments that require immediate access to information. To facilitate this fact of life, each chapter in this book was written as an entity to be as independent as possible from succeeding and preceding chapters. Thus, readers requiring access to specific information about a particular topic may be able to turn to the relevant chapter instead of reading all preceding chapters. However, an exception to this general rule might occur in instances where readers need background information concerning both the operation of a specific component of an optical transmission system and the use of an optical fiber transmission system within a particular operational environment. In this situation, it will be necessary to read at least two chapters: one chapter describing the operation of a particular optical transmission system component and the other chapter focusing on the operational environment for which the reader requires information.

Now that we have an appreciation of where this chapter is headed, let's commence our reading effort by focusing on the advantages associated with the use of optical transmission.

Advantages of Optical Transmission

When considering the use of an optical transmission system, most people think about the wide bandwidth of optical fiber. Although this is certainly important and provides the primary reason for many organizations migrating to optical transmission systems, it is not the only reason for considering the use of this type of transmission system. As we will note shortly, there are a number of reasons, some of which may be more applicable to one type of organization than to another type.

Table 1.2 lists seven key advantages associated with the use of a fiber-optic transmission system. Because many persons associate bandwidth primarily with optical fiber, let's first discuss this topic.

Bandwidth

As noted earlier, most persons rightly consider the wide bandwidth of optical fiber as the key advantage associated with an optical transmission system. Because the information transmission capacity of a communications system is directly proportional to bandwidth, the high frequency of light makes it possible to transmit data at extremely high rates. While gigabit data transmission rates are a common capability on optical fiber, research indicates that operating rates up to and beyond 10^{14} bits/s can be achieved on fiber-optic transmission systems. When compared to the 56-kbit/s rate on the public switched telephone network (PSTN) and the approximate 1-Mbit/s rate supported by an asymmetrical digital subscriber line (ADSL), it is easy to visualize how optical fiber provides many magnitudes of capacity enhancement over conventional copper cable. This explains why a thin glass fiber can easily transmit hundreds of thousands to millions of telephone conversations simultaneously. While this is certainly a valid reason for communications carriers migrating their backbone infrastructure to optical fiber, it also provides the rationale for the use of fiber in numerous other applications.

TABLE 1.2 Advantages of Fiber-Optic Transmission System

Large bandwidth permits high data transmission, which also supports the aggregation of voice, video, and data

Technological improvements are occurring rapidly, often permitting increased capacity over existing optical fiber

Immunity to electromagnetic interference reduces bit error rate and eliminates the need for shielding within or outside a building

Glass fiber has low attenuation, which permits extended cable transmission distance

Light as a transmission medium provides the ability for the use of optical fiber in dangerous environments

Optical fiber is difficult to tap, thus providing a higher degree of security than possible with copper wire

Light weight and small diameter of fiber permit high capacity through existing conduits

For example, within a building the bandwidth of fiber makes it suitable for multiplexing the transmission requirements of different tenants onto a common fiber for connection to a communications carrier. Within a floor in a building, fiber can be used to support the high-speed transmissions of Gigabit Ethernet. Thus, fiber is suitable for intrabuilding, interbuilding, national, and international transmission applications.

Technical Improvements

Another benefit associated with the use of optical networking devices is the pace of technological improvement occurring in this technology field. This progress far overshadows any effort at improving copper-based technology. For example, during 1995 equipment developers succeeded in creating tunable lasers capable of transmitting multiple bands of light through a single strand of optical fiber cable. Each band of light operates at a separate frequency and in effect provides multiple communication paths over or through a common fiber. The first generation of such systems, which is referred to as *wavelength division multiplexing* (WDM), provided four and eight separate optical frequencies or, as referred to in trade press articles, "split a fiber into four and eight colors." While this book was being written, several equipment vendors began marketing systems referred to as *dense wavelength division multiplexing* (DWDM that are capable of splitting a fiber into 128 colors. In addition, other vendors announced the development of prototype DWDM systems that have derived 1024 channels through a common fiber using a tunable laser. With just one optical frequency on a fiber providing several magnitudes of the capacity of a copper pair, it is quite easy to visualize the superiority of DWDM to the use of metallic pair wiring. Specifically, one strand of optical fiber is now capable of transporting the equivalent of a half-million T1 lines, with each of the latter capable of carrying 24 voice conversations. Thus, in terms of data transmission rate and capacity, a DWDM system can transport approximately 12 million voice calls!

While the capacity improvement afforded WDM and DWDM are considerable, it should also be mentioned that this improvement often occurs over an existing optical fiber infrastructure. Because the cost associated with the installation of optical fiber commonly accounts for most of the total cost of installing an optical transmission system, the ability to expand capacity by changing the transmitter and receiver represents a tremendous economic benefit. This also explains why communications carriers have literally been gobbling up WDM and DWDM components as rapidly as the manufacturers of such components can produce them.

Electromagnetic Immunity

In a copper cable environment the flow of electrons can be altered by electromagnetic interference (EMI). In an optical transmission system, light in the form of photons traverses down the fiber. Photons are not affected by electromagnetic interference, and there is no photonic equivalent. For instance, optical systems in a building are immune (i.e., impervious) to a noisy electrical environment resulting from EMI produced by machinery including devices such as an electrical light fluorescent ballast and an electronic pencil sharpener. In a WAN environment, protection is extended to obtain immunity to sunspots and other electrical disturbances.

The immunity from EMI permits optical fiber to be routed within a building without having to worry about the location of sources of electromagnetic radiation, such as fluorescent ballasts. In addition, immunity from EMI optical cable avoids one of the most common causes of transmission errors. Thus, you can normally expect a fiber-optic transmission system to have a bit error rate considerably lower than that achievable with a copper-cable-based system.

Another advantage of the immunity of optical fiber from EMI is the fact that multiple cables in close proximity to one another do not generate crosstalk as do copper cables. This means that multiple fibers can be bundled together without requiring special shielding and provides another option for the use of an optical transmission system.

Low-Signal Attenuation

In a copper cable signal, attenuation is directly proportional to frequency, with high frequencies attenuating more rapidly than lower frequencies. In comparison, the flow of light in the form of photons in an optical fiber does not exhibit the characteristics described above, and the attenuation in the signal is relatively independent of the frequency.

In comparison to copper-based transmission systems, the lack of signal loss at high frequencies permits fiber-optic systems to transmit information for longer distances before requiring the signal to be amplified.

Environment Utilization

In a copper-based system, the flow of electrons can result in a spark or shock. This means, for example, that the routing of copper cable in an oil refinery or grain elevator environment could result in an explosion if the wires short-circuit. In comparison, the routing of optical fiber through such locations prevents the risk of ignition caused by faulty wiring as light, instead of electrons, now circulates through the dangerous environment.

While only a small percentage of optical fiber may be used in refineries, chemical plants, and grain elevators, the use of optical fiber in an office environment is now commonly encountered and also eliminates the potential for electrical hazards. For example, when used in a building, fiber provides complete electrical isolation between a transmitter and a receiver. This means that the common ground between these two components that is required with the use of copper conductors is eliminated.

Another key advantage of optical fiber in many locations is the fact that you can route this type of fiber in ceilings or under floor panels without having to run the cable through a conduit. When this author first performed ceiling routing of optical fiber without using a conduit many years ago, the fire marshal appeared to have a high degree of anxiety as the author explained to him that light, and not electricity, flowed through the cable. Now, in this more modern era, many building codes have been revised to reflect the fact that optical fiber does not and cannot transport electricity. When complying with modern building codes, the ability to route fiber without the need for installing a conduit in the form of 200 to 300 ft of metal pipe can easily save several thousand dollars.

Security

If you watch spy movies you might recall one or more scenes when a person in a van parked outside of a fence around a high-security building points an antenna toward the building, turns some dials, and watches a monitor that displays a message being typed by the occupant of the building. What the "spy" in the van is doing is operating a directional antenna that has sensitive electronics attached that receive electronic radiation emitted by electronic equipment inside the building.

All electronic equipment and transmission systems radiate energy. By "reading" the radiated energy it becomes possible to note information being transmitted or received. Of course, what the movie may not show nor the story line tell is the fact that a spy using such a van in front of the CIA (U.S. Central Intelligence Agency) headquarters might be a bit noticeable, especially since the van contains a lot of electronics that require a significant amount of power. In addition, it is rather difficult to hide the directional antenna on top of the vehicle.

The electronic emissions are referred to in the spy (espionage) community as *tempest*. To make a building or another site "tempestproof," the site is hardened by lead shielding. When copper cable is used to connect buildings, these cables, too, must be shielded. In comparison, an optical fiber does not radiate energy and thus does not require shielding.

Another benefit of optical fiber with respect to security is the fact that, unlike the tapping of a copper circuit, which could be unnoticeable, tapping an optical fiber requires the insertion of an optical splitter. This insertion results first in a loss of signal and then in a loss of signal strength. Thus, it is easier, with applicable equipment, to note a tap of an optical signal than an electrical signal. In light of the preceding observations, you might make a reasonable guess as to the type of cable favored for use by intelligence agencies to interconnect buildings.

Weight and Size

The glass on plastic used in an optical fiber cable is a thin strand of material. Even when fiber-optic cable is surrounded by a jacket for protection, the weight and diameter of the resulting cable are considerably less per meter than those of copper cables, which provide only a fraction of the transmission capacity of optical fiber.

Disadvantages of Optical Fiber

Although we noted a considerable number of advantages associated with the use of an optical transmission system, we should also note some of the limitations associated with the use of this type of transmission system. Two of the key limitations associated with the use of optical fibers are cable splicing and the cost of optical fiber.

Cable Splicing

When you need to extend a copper cable within a building, it is possible to simply strip the insulators on the pair to be extended and strip another wire pair, twist the wire strips for each pair together, and bond them with electrical tape. In so doing, your primary concern is to ensure that each wire pair is correctly mated to the wire on the extension. When you splice optical fiber, you must align the center core of one fiber to another, a much more difficult procedure. In addition, your options for joining fibers include welding or fusing, gluing, or the use of mechanical connectors. Each method is more time-consuming than simply twisting electrical conductors together and can result in an increase in cabling cost. This is especially true because personnel normally require training to splice optical fiber to minimize optical loss, which can adversely affect the transmission capability of a fiber.

Welding or fusing normally results in the least loss of transmission between splice elements. However, you must clean each fiber end, then align and carefully fuse the ends using an electric arc. This is time-consuming but can result in the least amount of signal loss between joined elements. Gluing or an epoxy method of splicing requires the use of bonding material that matches the refractive index of the core of the fiber. Thus, you just can't run out to the nearest CVS, Kmart, Office Depot, or another store and use any type of glue. Again, gluing is time-consuming and results in a higher loss of signal power than does fusing.

A third method available to join fiber is the use of connectors. Although mechanical connectors considerably facilitate the joining of fibers, they result in more signal loss than do the other two methods and can reduce the span of the fiber to a smaller distance.

Fiber Cost

A second limitation associated with an optical transmission system is the cost of optical fiber. While you can compute the cost of fiber on a (bit/s)/km basis, which will always be less than that for copper cable, when used within a building, some organizations may require only a fraction of the capacity of the optical fiber. For this reason, it is often difficult to justify fiber to the desktop and similar applications where the cost of copper cable, such as category 5 cable, may be half or less than the cost of fiber. Now that we have an appreciation for the advantages and disadvantages associated with optical transmission systems, we will conclude this chapter by previewing the succeeding chapters in this book. As noted earlier in this chapter, you can use this information by itself or in conjunction with the index of this book to locate specific information if you wish to turn to a topic of immediate interest.

Chapter Preview

In this section we will preview the remaining chapters in this book. Although chapters were written to be as independent from one another as possible, readers new to this field should read chapters in the sequence presented. With that said, let's take our minitour of those chapters.

Understanding Light

In Chapter 2 we focus, no pun intended, on developing an appreciation for a phenomenon that we take for granted and of which we see only a portion. That phenomenon is light. In Chapter 2 we will examine the frequency spectrum, noting the different bands in the spectrum such visible light, ultraviolet light, gamma rays, and other bands that students usually first became acquainted with during high school or college physics. We will also review a variety of terms associated with light and will discuss power measurements that are important for determining the performance level of an optical transmission system.

Understanding Fiber

For light to act as an optical transmission system, it must flow on a medium. That medium is fiber, which is the topic of Chapter 3. In that chapter we will discuss the different types of fiber that can be used in an optical transmission system, how light flows down and through each type of fiber, and the advantages and disadvantages associated with the use of each type of fiber. Because it is important to understand the transmission characteristics of light as it flows into and through a fiber, we will also examine many terms associated with light distribution in a fiber. Other topics covered in Chapter 3 include a brief introduction to the increased transmission capacity of a fiber through wavelength division multiplexing (WDM) and dense wavelength division multiplexing (DWDM).

Light Sources and Detectors

Like any type of transmission system, an optical transmission system requires a transmitter and a receiver. The *transmitter* is the light source; the receiver is an optical detector, also referred to as a *photodetector*. Thus, in Chapter 4 we will describe the operation of light-emitting diodes (LEDs) and lasers as well as different types of optical detectors.

Fiber in the LAN

Beginning in Chapter 5, we discuss applications that use optical transmission systems. In that chapter we will examine the use of optical transmission in the local area network, investigating how one version of Fast Ethernet and several versions of Gigabit Ethernet are dependent on the use of fiber. In addition, we will also summarize the functions of the fiber channel, which enables high-speed access between computers and data storage and represents the cornerstone of storage area network communications.

Fiber in the WAN

Once we discuss the practical use of fiber in the local area network, we will focus on the wide area network. In Chapter 6 we will discuss and describe the Synchronous Optical Network (SONET) and its European counterpart, the Synchronous Digital Hierarchy (SDH). We will also examine WDM and DWDM in more detail and an evolving technology referred to as the *Internet Protocol* (IP) over SONET.

Fiber in the Neighborhood

The growth in the use of digital subscriber line (DSL) and cable modems requires telephone companies and cable operators to route fiber into neighborhoods to satisfy the increased transmission requirements of subscribers. Thus, in Chapter 7 we will examine the technology that we can collectively refer to as *fiber to the neighborhood.*

Fiber in the Building

No book covering optical transmission components would be complete without discussing the use of fiber within a building. In Chapter 8, our concluding chapter, we will examine the use of fiber modems, fiber multiplexers, and even WDM within a building.

CHAPTER 2

Understanding Light

Information is transmitted over different types of fiber by means of light. If we compare fiber to a copper conductor, we can note several similarities between the two. In a fiber environment, a light source replaces the electronic transmitter that generates pulses and is used by a copper-based digital transmission system, while the conductor is a glass or plastic fiber in place of a twisted-pair wire or coaxial-cable conductor. Because of the central role of light in a fiber-optic transmission system, it is important to inquire into what we normally take for granted.

In this chapter we will literally focus our eyes and attention on light. In so doing, we will discuss how light can be described in terms of particles and waves, and where different colors are located in the frequency spectrum. As we discuss light we will introduce several terms that may appear new to many readers while serving as a refresher for others. Because this book is oriented toward communications applications, we will also discuss such terms as *bandwidth, signaling rates, power measurements,* and *channel capacity.* One or more topics covered in this chapter may leave you scratching our head in an attempt to determine how the topic relates to communications over a fiber. To paraphrase Indiana Jones, this author will say, "Trust me."

Describing Light

If you previously took a course covering physics, you probably read a chapter covering light. Although this may have been illuminating (no pun intended), you probably noted that light can be described as a particle or as a wave, which in itself is a somewhat obscure idea.

Light as a Particle

If you consider light to represent a flow of particles, each particle is referred to as a photon (from the Greek word *photos,* meaning light). A photon has only energy and no mass; hence a beam of light consists of a flow of photons but no mass. The intensity of the beam is then directly proportional to the flow of photons; thus a higher-intensity beam has a greater flow of photons.

Describing light as a flow of photons makes it relatively easy to visualize its absorption. In fact, this is what Albert Einstein did in 1905, to describe what is now referred to as the *photoelectric effect.* Dr. Einstein focused light onto a metal surface that was placed in a vacuum. At the same time he placed an electron detector above the metal to determine whether photons transferred their energy to electrons as the photons were absorbed by the metal. Dr. Einstein's photoelectric effect is illustrated in Figure 2.1 and represents the basis for demonstrating the particle flow of light.

Light as an Electromagnetic Wave

Alternatively, light can be described as a wave. In the wonderful world of communications, it makes more sense to describe data transmission in terms of waves rather than individual particles, because nobody talks about the transmission of information over a wire circuit as a flow of electrons. Instead, we discuss how electromagnetic waves are modulated to convey information. As we proceed through this book, we can describe and discuss the flow of light through a fiber conductor without having to refer to the particle nature of light. Thus, in the remainder of this book we will describe and discuss light primarily in terms of an electromagnetic wave.

An *electromagnetic wave* can be considered as a continuum of oscillating electric and magnetic fields moving in a straight line at a constant velocity. For light, that velocity is approximately 186,000 miles per second (mi/s), or in the metric [Système International (SI)] system of measure-

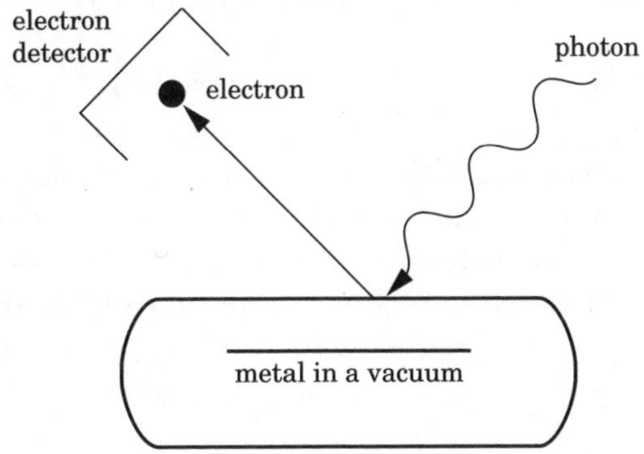

Figure 2.1
The photoelectric effect illustrates the transfer of light energy to an electron and demonstrates the particle nature of light.

ment, 300,000,000 meters per second (m/s). As a Trivial Pursuit note, although scientists long considered the speed of light in a vacuum to represent the highest rate at which anything can travel, in July 2000 this changed. In an experiment conducted in Princeton, New Jersey, physicists transmitted a pulse of laser light through a chamber filled with cesium vapor so quickly that it left the chamber before fully entering it. According to press reports, the pulse traveled 310 times the distance it would have traversed if the chamber had been a vacuum. Although a practical application of this experiment may not be developed for quite some time, this observation does indicate that the speed of light can be pushed beyond previously assumed boundaries.

NEWTON'S PRISM Although people viewed rainbows a long time prior to the birth of Isaac Newton, Newton's work in 1672 formed the basis for the modern treatment of light as a wave. During 1672, Newton discovered, in experiments with a prism, that light could be split into a series of colors. As he used a prism to analyze light, Newton noted that the colors produced by light passing through the prism were arranged in a precise order; specifically, red was followed by orange, yellow, green, blue, indigo, and violet. If you remember the good old days when you pulled an all-nighter to study for a physics test, the name Roy G. Biv may come to mind. That fictional name was used by many students as a way to remember the order of colors in the light spectrum.

Although Newton's prism was the pioneering effort in defining the components of visible light, it took many years of effort to further clarify the color of objects. When modern science was applied to differentiate colors, it was concluded that no single wavelength exists for many colors, which are created by a mixture of wavelengths. For example, purple represents a mixture of red and violet wavelengths.

MAXWELL'S EFFORT Approximately 100 years after Newton's work, James Clerk Maxwell showed that light was a form of electromagnetic radiation. Maxwell, a nineteenth-century physicist, had a keen interest in electricity and magnetism. One of his major accomplishments was the development of mathematical equations that describe how electricity and magnetism work together to produce light and radio waves. He also developed a color triangle which is commonly used in art and physics classes to explain the relationship between the primary colors (red, blue, and green) to other colors. Later in this chapter we will examine the Maxwell color triangle. Work by Maxwell and other scientists showed that light could be categorized as a series of electromagnetic waves, with the human eye responding to different wavelengths, referred to as *visible light,* while ignoring other wavelengths.

Basics of Electromagnetic Waves

When the human eye views light as a series of electromagnetic waves, it becomes a relatively simple task to place this light in the frequency spectrum and to discuss its ability to convey information. However, before we do this, let's describe and discuss the basic parameters of a wave: its frequency and wavelength.

Frequency

In a series of electromagnetic waves, represented as light, the waves oscillate at different frequencies. Here the term *frequency* is used to refer to the number of periodic oscillations or waves that occur per unit time.

Figure 2.2 illustrates two oscillating waves at different frequencies. The top portion of the figure illustrates a sine wave operating at one cycle per second (cps). Note that the term *cycles per second* in general has been replaced by the synonymous term *hertz* (Hz), after Heinrich Hertz, a famous German scientist known for his work in the field of electromagnetism during the latter part of the nineteenth century. The lower portion of Figure 2.2 shows the same sine wave after its oscillating rate has been doubled to 2 Hz.

The time required for a signal to be transmitted over a distance of one wavelength is referred to as the *period* of the signal. The period represents the duration of the cycle and can be expressed as a fraction of the frequency. Thus, if T represents the period of a signal and f is the signal's frequency, then $T = 1/f$. We can also express frequency in terms of the period of a signal: $f = 1/T$.

In Figure 2.2 we note that the sine wave whose period is 1 s has a frequency of 1/1 or 1 Hz. Similarly, the second sine wave whose period is 0.5

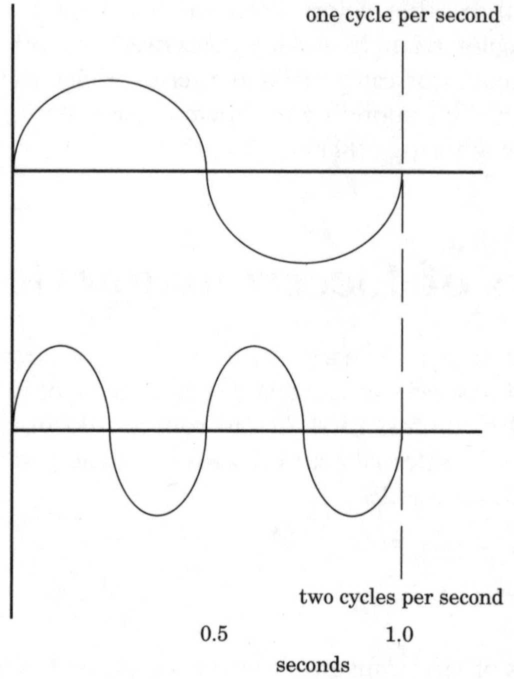

Figure 2.2 Oscillating sine waves at different frequencies.

s has a frequency of 1/0.5, or 2 Hz. The period and the frequency of a signal are inversely proportional to one another. Thus, as the period of a signal decreases, its frequency increases.

Wavelength

The period of an oscillating signal is also referred to as the signal's *wavelength,* denoted by the Greek lowercase symbol lambda (λ). The wavelength of a signal can be obtained by dividing the speed of light (3×10^8 m/s) by the signal's frequency in hertz. In actuality, the speed of light is 299,792,458 m/s. However, since the goal of this book is to explain the transmission of information over fiber optics and not to teach physics, we will round the speed of light to 3×10^8 m/s. Thus, we can denote the wavelength of a signal for either radio or light waves as follows:

$$\lambda\ (\text{m}) = \frac{3 \times 10^8}{f(\text{Hz})}$$

Note that you can adjust the numerator and denominator of this equation. In so doing, you can adjust the frequency from hertz (Hz) to kilohertz (kHz), megahertz (MHz), gigahertz (GHz), and terahertz (THz). As a refresher for those of you who may be a bit rusty remembering prefixes for the power of 10, Table 2.1 lists nine common prefixes and their meanings. Thus, the use of GHz represents billions of cycles per second, while MHz refers to millions of cycles per second, and kHz is used to refer to thousands of cycles per second.

Wavelength can be expressed in terms of Hz, kHz, MHz, GHz, and THz as follows:

$$\lambda\,(\mathrm{m}) = \frac{3 \times 10^8}{f(\mathrm{Hz})} = \frac{3 \times 10^5}{f(\mathrm{KHz})} = \frac{300}{f(\mathrm{MHz})} = \frac{0.3}{f(\mathrm{GHz})} = \frac{0.0003}{f(\mathrm{THz})}$$

Because wavelength is expressed in terms of the speed of light divided by the frequency, the frequency can be defined in terms of the speed of light divided by the wavelength. This gives

$$f(\mathrm{Hz}) = \frac{3 \times 10^8}{\lambda\,(\mathrm{m})}$$

Just as we computed the wavelength in terms of varying frequency, we can also compute frequency in terms of varying speed of the light

TABLE 2.1
Common Prefixes of the Powers of 10

Prefix	Meaning	
pico	1/1,000,000,000,000	trillionth
nano	1/1,000,000,000	billionth
micro	1/1,000,000	millionth
milli	1/1000	thousandth
kilo	1000	thousand
mega	1,000,000	million
giga	1,000,000,000	billion
tera	1,000,000,000,000	trillion
peta	1,000,000,000,000,000	quadrillion

constant. Thus, we can compute frequency in terms of a wavelength in meters as follows:

$$f(\text{Hz}) = \frac{3 \times 108}{\lambda\ (\text{m})}$$

$$f(\text{kHz}) = \frac{3 \times 105}{\lambda\ (\text{m})}$$

$$f(\text{MHz}) = \frac{300}{\lambda\ (\text{m})}$$

$$f(\text{GHz}) = \frac{0.3}{\lambda\ (\text{m})}$$

$$f(\text{THz}) = \frac{0.0003}{\lambda\ (\text{m})}$$

Although we normally want to be precise in the computations we make, in the world of communications there are several rules of thumb that can be used to expedite computations. For metric computations, you can estimate the wavelength in centimeters as follows:

$$\lambda\ (\text{cm}) = \frac{30}{f(\text{GHz})}$$

The Frequency Spectrum

Using our preceding discussion of frequency and wavelength, let's look once more at the electromagnetic radiation spectrum, which is more generally referred to as the *frequency spectrum.* By doing this we can better comprehend the relationship between light and other types of electromagnetic waves, such as audible conversations, AM and FM (amplitude and frequency modulation) radio, different types of television broadcasts, and microwave and infrared communications.

Figure 2.3 illustrates, among other things, the locations of popular types of communication in the known frequency spectrum. The left side of Figure 2.3 indicates the frequency in powers of hertz; the right side, the wavelength in terms of fractions of a meter. If you carefully examine Figure 2.3, you will note that visible light appears in the micrometer (μm) wavelength region, where red light has a wavelength of 0.68 μm. In many physics books the wavelength of visible light is listed in terms of nanometers (nm), where one nanometer equals one thousand millionths,

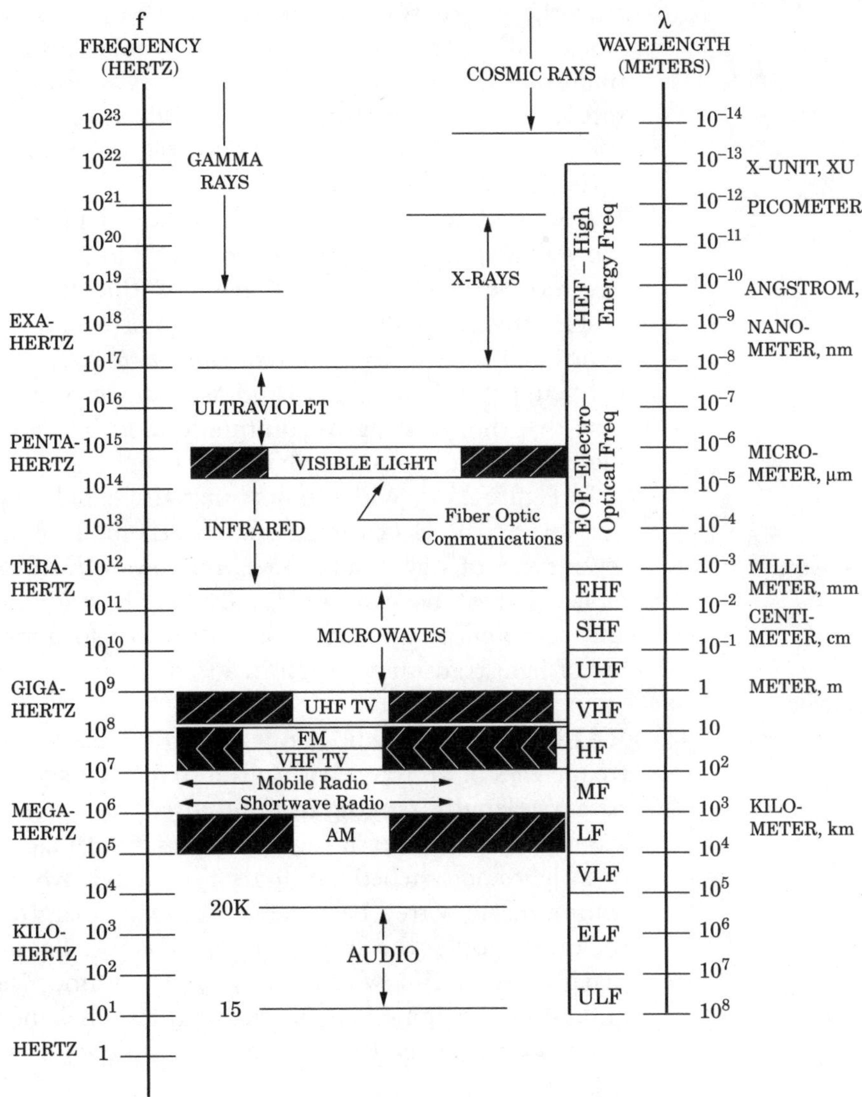

Figure 2.3
The known frequency spectrum.

or a billionth, of meters (10^{-9} m). Thus, red light is normally referred to as having a wavelength of 680 nm.

In examining the frequency spectrum shown in Figure 2.3, note that the rainbow of colors we're able to see represents only a very small portion of the electromagnetic spectrum. At the low end of the spectrum are radio waves that have wavelengths billions of times longer than those of visible light. In fact, visible-light wavelengths are considerably smaller than the thickness of a human hair, whose diameters might be equivalent to 100 visible-light wavelengths. While visible light has a relatively small wavelength, X rays and gamma rays are even shorter in length and, similar to longer wavelengths, are invisible to the naked eye. In fact, radiation with wavelengths shorter than 400 nm or longer than 700 nm are invisible to the naked eye.

If we look very closely at the right side of Figure 2.3, we will note that the wavelength from approximately 10^{-3} to 10^{-8} m is referred to as the electrooptical frequency (EOF) band. Within this band are infrared and visible light. We will also note a tiny horizontal line below the visible-light band labeled *fiber-optic communications.* When we discuss the use of light-emitting diodes (LEDs) and laser diodes later in this book, we will note that the wavelengths commonly used by those light generators are slightly above the wavelength of visible light.

In Figure 2.3 we will also note that visible light represents only a small portion of the electromagnetic spectrum. That spectrum consists of seven types of waves: radio waves, microwave, infrared, visible light, ultraviolet, X rays, and gamma rays. By briefly studying each type of wave and its potential utilization, we can begin to understand the potential use of light transmission within a fiber.

RADIO WAVES Radio waves have the longest wavelength and lowest frequencies of all waves within the frequency spectrum. Such waves can have wavelengths ranging from hundreds of meters at low frequencies to less than a centimeter at higher frequencies. In fact, if you stayed up late one night and watched a military-type movie where a submarine had to communicate with a base station while submerged, you probably heard the captain order the antenna to be deployed. Because a submerged submarine uses very-low-frequency (VLF) communications, the resulting radio waves are quite long, requiring the submarine to deploy a spool of wire that can extend a distance of a mile or more to function as an antenna.

As we move up the frequency spectrum we encounter AM, VHF TV, FM, and UHF (ultra-high-frequency) TV. AM radio uses the 550- to 1650-kHz band, while FM radio occurs in the 88- to 108-MHz band. Concerning television, in the United States VHF TV resides in the 54- to 216-MHz band; UHF TV, in the 220- to 500-MHz band. Although not shown in Figure 2.3, wireless communications that occur in the 800- and 1900-MHz bands results in relatively short wavelengths, which explains why the handset in your purse or connected to your belt has a relatively short antenna.

MICROWAVES Microwaves are shorter than radio waves but longer than infrared. Two of the more popular uses for microwaves are for cooking and radar.

Because water and fat molecules absorb energy in the form of waves generated at approximately 2.45 GHz, a microwave oven generates waves at that frequency. To do this, the oven uses a small metal box with several ridges that are sealed in a vacuum to generate electromagnetic waves at a frequency of 2.45 GHz. That box, referred to as a *magnetron,* represents the most costly component in a microwave oven. A second common use for microwaves occurs at frequencies around 10 GHz. Microwaves at that frequency are used for radar, where reflected microwaves directed at a specific target are returned at a shifted frequency. That returned frequency, which results from the Doppler effect, enables the speed of the object to be determined, to the chagrin of many motorists.

INFRARED Infrared waves occur in that portion of the frequency spectrum above the microwave region but below visible light. Infrared rays are generated by hot objects in the form of infrared radiation, which explains how devices with sensors tuned to the infrared region can detect cars, tanks, and other vehicles and pinpoint the loss of heat from homes and apartments. Another use for infrared is for short-distance communications; for instance, many laptop and notebook computers have an infrared sensor.

VISIBLE LIGHT As mentioned earlier, visible light resides in the electromagnetic spectrum with wavelengths between 400 and 700 billionths of a meter (400 to 700 nm). To better appreciate its positioning in the frequency spectrum, refer to Figure 2.4, which provides an "exploded" view of the frequency spectrum indicating the approximate wavelengths for red and violet light, which represent the boundaries of visible light. Note that red light has a wavelength of 680 nm while violet light has a wavelength of 410 nm.

Because every electromagnetic wave has a unique frequency and wavelength associated with it, we can draw the wave corresponding to each color of light. For example, Figure 2.5 illustrates the electromagnetic wave corresponding to the color red. Note that its frequency is 428.570 GHz, or 428,570 billion cycles per second. Thus, when you view red light, your eye receives over 400 trillion waves per second. Also note that the wavelength of red light is 700 nm long, which is 7 ten-millionths of a meter.

When we discuss the operation of lasers later in this book, we'll see that the wavelength employed varies with the application as well as the properties of optical fiber. For local area networks (LANs), lasers typically operate at wavelengths of 850 and 1300 nm. For cable TV and telephone communications, lasers typically have source wavelengths of 1310 and 1550 nm. One of the key properties of optical fiber that governs the use of lasers at certain frequencies is the relationship between attenuation and wavelength when light flows through a fiber-optic system. Figure 2.6 provides a general illustration of that relationship. Note that one of the major operational requirements of an optical fiber transmission system is to provide the lowest possible level of attenuation when light is being transmitted. If you examine Figure 2.6, you will note three "windows" where attenuation at certain wavelengths are generally lower than at other wavelengths. By using lasers or LEDs tuned to wavelengths within these windows, it becomes possible to transmit for longer distances before requiring the use of amplifiers. This, in turn, reduces the cost of

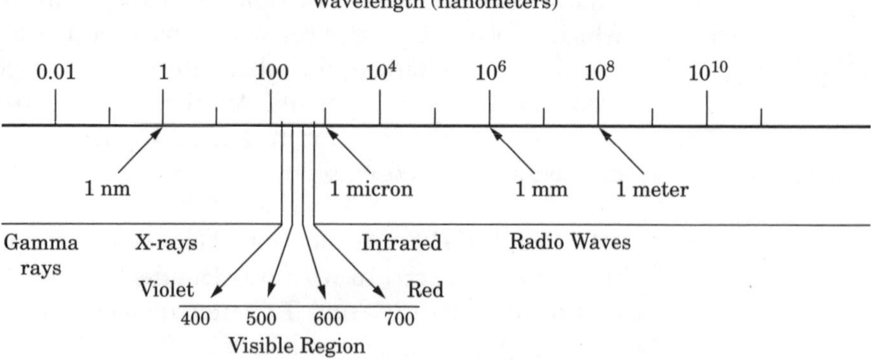

Figure 2.4
An exploded view of the frequency spectrum centered on the visible-light region.

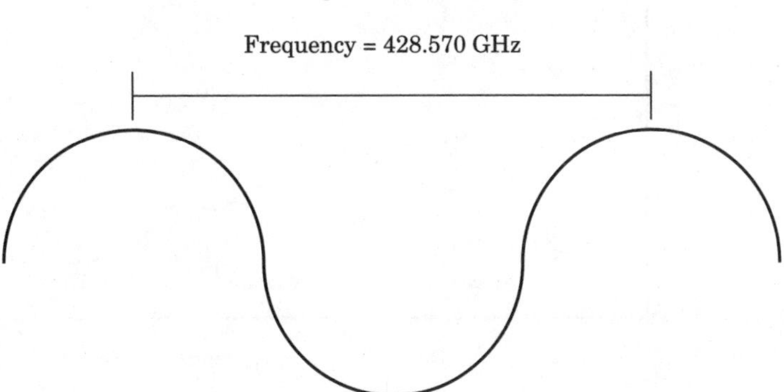

Figure 2.5 The electromagnetic wave corresponding to the color red.

the fiber-optic transmission system since it reduces the number of amplifiers required. Because we need to understand some additional terms, we will revisit the transmission of light in optical fiber to include a more detailed examination of the relationship between attenuation and the wavelength of light in Chapter 3 when we discuss optical fiber.

ENERGY AND LIGHT If we again think back to days spent in a physics class, we might remember using the term *electronvolt* (eV) to describe visible light when we characterized light as a stream of photons. As a refresher, the electronvolt represents the energy gained by an electron that passes across a positive voltage of one volt (V). Thus, an electron moving from a negative metal plate to a positive metal plate when current flows in a common 1.5-V C battery would result in 1.5 eV of energy.

Returning to the use of electronvolts, visible light in the form of photons has an energy range of 1.8 to 3.1 eV; these electronvolts are visible as that energy range triggers the photo receptors in a human eye. We can also note visual perception in terms of wavelength. Thus, the electromagnetic spectrum that can be seen by the naked human eye (macroscopically) ranges from approximately 4.3×10^{14} Hz for red to 7.5×10^{14} Hz for violet. Lower energies that have longer wavelengths are invisible to the naked human eye but can be detected by special sensors or even by the human ear. For example, infrared, television, and radio waves can be detected by equipment, while sound frequencies of up to 20 kHz can be detected by the human ear. Electromagnetic waves with energies above 3.1 eV and shorter wavelengths are also undetected by the unaided human eye; however, they can be detected by different types of equipment. For example, X rays can be detected by lead plates or photographic film.

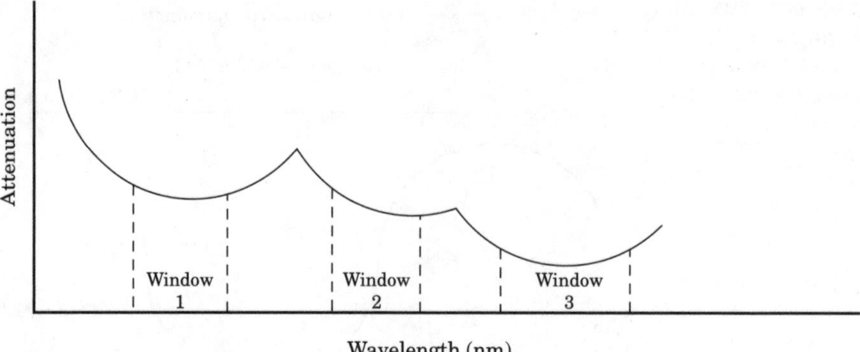

Figure 2.6 The general relationship between attenuation and the wavelength of light results in three "windows" where a particular group of wavelengths encounter the least amount of resistance.

ENERGY AND WAVELENGTH As the energy of photons increases, the wavelength of these photons decreases; thus, the energy of light is inversely proportional to its wavelength. For example, red-light photons have an energy level of approximately 1.8 eV and a wavelength of 700 nm, while violet has an energy level of 3.1 eV and a wavelength of 400 nm. We can also note that the energy of photons is then proportional to the light frequency, such as $E = hf$, where h represents a famous constant that we will note shortly. This equation was first proposed by Max Planck when he investigated the emission of radiation by heated solids. The constant h has the value 4.136×10^{-15} eV/s, and is referred to as the *Planck constant,* while f represents the frequency of the photon.

We can also express the energy of a photon in terms of its wavelength and the speed of light in a vacuum. Ignoring intermediate steps and following the expression of Indiana Jones ("Trust me"), we can rewrite Planck's equation as follows:

$$E = \frac{hc}{\lambda}$$

Because hc has the value 1240 eV/nm, we can rewrite Planck's equation further as follows:

$$E = \frac{1240}{\lambda\ (\text{nm})}$$

On the basis of these observations, we can plot the relationship between the energy in electronvolts and wavelengths for photons in the visible region of the electromagnetic spectrum (see Figure 2.7).

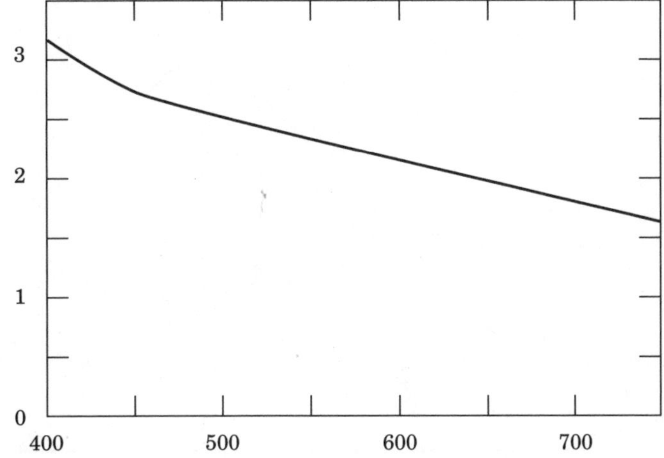

Figure 2.7
The relationship between energy (in eV) and wavelength (in nm) for the visible-light spectrum.

COLOR When we consider the components of visible light, our good friend, Roy G. Biv, tells us that there are seven distinct colors. Thus, a logical question you might have, unless you were an art student or a physics major, is how the human eye perceives other colors.

The human eye has three receptors that respond to different wavelengths. In fact, a color triangle popularly used in art and physics classes and which is attributed to the work of James Clerk Maxwell can be used to explain how humans view color. Let's now consider the Maxwell color triangle, which is illustrated in Figure 2.8.

Note that the three primary colors (red, blue, and green) are located at the three apices. Most, but not all, colors can be produced by mixing the three primary colors.

Maxwell's work paved the way for a more sophisticated color triangle now used to represent all colors. That triangle was developed by the Commission Internationale de l'Éclairage (CIE) in 1931 and postulates three ideal, but nonexistent, primary colors, one located in each corner of the triangle. This forms a display of the amount of ideal blue, red, and green that would have to be mixed to form different colors. This also explains why modern videodisplays use blue, red, and green light sources to generate color.

ULTRAVIOLET In an era where many people are concerned about ozone depletion, ultraviolet radiation has received a significant amount of publicity. Most of us associate ultraviolet radiation with sunburn and

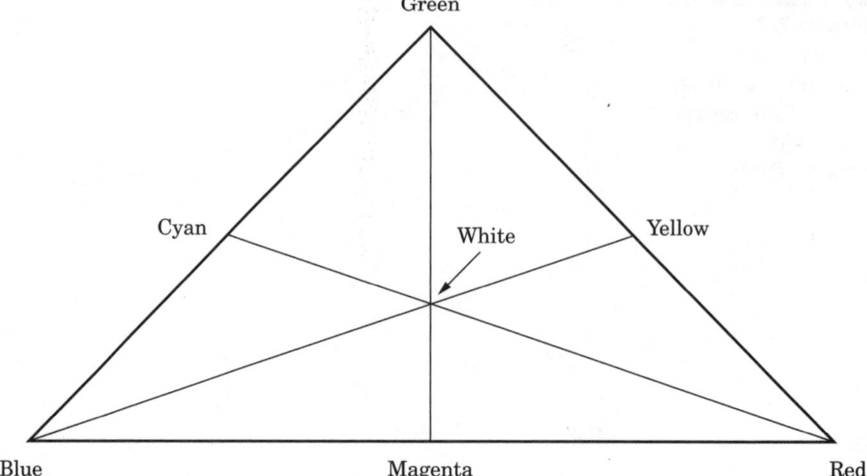

Figure 2.8
The color triangle, which is attributed to James Clerk Maxwell, illustrates how the three receptors in the human eye respond to different wavelengths.

the potential occurrence of skin cancer. Technically, ultraviolet radiation represents that portion of the electromagnetic spectrum above visible light. Thus, ultraviolet radiation has a shorter wavelength and higher frequency than does visible light.

The wavelength of ultraviolet radiation ranges from approximately 400 billionths of a meter to about 10 billionths of a meter. Ultraviolet waves are generated by the sun, and the ozone layer blocks most of those rays. This explains why the depletion of the ozone layer in the earth's atmosphere is a great matter of concern. Another source of ultraviolet radiation is ultraviolet lamps that many persons associate with tanning salons. However, ultraviolet light is also used in hospitals to treat skin conditions and kill different types of bacteria.

X RAYS As we move up the ladder of the frequency spectrum, we find high-energy waves, referred to as *X rays,* residing above the ultraviolet band. X rays were discovered by Konrad Röntgen in 1895 when he determined, by accident, that a fluorescent plate glowed even when it was far away from a tube conveying an electron beam. When he placed his hand in front of the fluorescent plate, he was able to see the bones in his hand; this discovery paved the way for the use of X rays in such diverse applications as inspecting welds in pipes and checking body parts.

The wavelength of X rays ranges from approximately 10 billionths of a meter to about 10 trillionths of a meter. Thus, X rays are electromagnetic waves with shorter wavelengths than those of ultraviolet light.

GAMMA RAYS Located toward the top of the known frequency spectrum, gamma rays have wavelengths under 10 trillionths of a meter. Because of their high frequency, gamma rays are more penetrating than X rays.

Gamma rays are created by such radioactive substances (isotopes) as cobalt-60 (^{60}Co) and cesium-137 (^{137}Cs) and are also a side effect of nuclear explosions. When controlled, gamma rays are used to destroy cancer cells and are also used as a tracer when placed inside a patient's body. The latter represents a fantastic voyage quite similar to the movie by that name but without the need to miniaturize humans.

Power Measurements

As we have noted several times, the flow of light through a fiber-optic waveguide represents a communications system similar to copper cable in many respects; however, instead of electrons flowing down a circuit, we now have photons. Although the type of particles and conductors changed, we still use certain basic measurements to calculate the ratio of the received power to the transmitted power, regardless of the type of transmitter and transmission medium used for communications. In this section we consider two common power measurements: the bel (B) and the decibel (dB). We will use those measurements to discuss one of the more important metrics in communications, used to categorize the quality of transmission. That metric is known as the *signal-to-noise ratio.*

Bel

The *bel* represents the first in a series of power ratios developed to quantify the proportion of power between two points. As you might surmise, the bel owes its name to Alexander Graham Bell, the inventor of the telephone. The bel uses logarithms to the base 10 to express the ratio of power transmitted to power received. The resulting gain or loss for a circuit is given by the following formula:

$$B = \log_{10} \frac{P_0}{P_1}$$

where B is the power ratio in bels, P_0 is the power output or received, and P_1 is the input or transmitted power. The bel used logarithms to measure power because humans hear logarithmically and the first series of power measurements were applied to telephone circuits that carried audio. In other words, the human ear perceives sound or loudness on a logarithmic scale. For example, if we estimate the loudness of a signal to have doubled, the transmission power will have actually increased by a factor of approximately 10. Another reason for the use of logarithms in power measurements is that signal boosts (gain) due to amplification or signal loss due to resistance are additive. The ability to add and subtract when performing power measurements based on a log (logarithmic) scale simplifies computations. For example, a 10-B signal that encounters a 5-B loss and is then passed through a 15-B amplifier results in a signal strength of 10 −5 +15, or 20 B.

For those of you who are a bit rusty when it comes to logarithms, note that the logarithm to the base 10 ($\log_{10}$) of a number is equivalent to how many times 10 is raised to a power equal to that number. For example, $\log_{10}$ 100 is 2, $\log_{10}$ 1000 is 3, and so on. Because output or received power is normally less than input or transmitted power, the denominator in the preceding equation is normally larger than the numerator. To simplify computations, we can note a second important property of logarithms:

$$\log_{10} \frac{1}{X} = -\log_{10} X$$

As an example of how to use the bel to compute the ratio of power received to power transmitted, let's assume that received power was one-hundredth of the transmitted power:

$$\mathrm{B} = \log_{10}\left(\frac{1}{100/1}\right) = \log_{10} \frac{1}{100}$$

Because $\log_{10} 1/X = -\log_{10} X$, we obtain

$$\mathrm{B} = -\log_{10} 100 = -2$$

Note that a negative value indicates a power loss while a positive value would indicate a power gain. Although the bel was used for many years to categorize the quality of transmission on a circuit, industry required a more precise measurement. This resulted in the adoption of the decibel (dB) as a preferred power measurement.

Decibel

The *decibel* (dB) represents the standard method used today to denote power gains and losses. It is a more precise measurement as it represents one-tenth of a bel. The power measurement in decibels is computed as follows:

$$\text{dB} = 10 \log_{10}\left(\frac{P_O}{P_I}\right)$$

where dB is the power ratio in decibels, P_O is the output or received power, and P_I is the input or transmitted power. Returning to our previous example where the received power was measured to be one-hundredth of the transmitted power, the power ratio in decibels becomes

$$\text{dB} = \log_{10}\left(\frac{1}{101/1}\right) = \log_{10}\frac{1}{100}$$

Because $\log_{10} 1/X = -\log_{10} X$, we obtain

$$\text{dB} = -\log_{10} 100 = -20$$

DECIBELS ABOVE 1 mW Note that the terms *bel* (B) and *decibel* (dB) do not denote power values. Instead, they represent a ratio or comparison between two power values, such as input and output power. Because it is often desirable to express power levels with respect to a fixed reference, it is common to use a one-milliwatt (mW) standard input for comparison purposes. In the wonderful world of communications testing of copper cable in North America, 1-mW signal occurs at a frequency of 800 Hz. To remind ourselves that the resulting power measurement occurred with respect to a 1-mW input signal, we use the term *decibel-milliwatt* (dBm):

$$\text{dBm} = \frac{\text{output power}}{\text{1-mW input}}$$

and serves to remind us that the output power measurement occurred with respect to a 1-mW test tone. Although the dBm is referred to as *decibel-milliwatt* in most literature, it actually means one decibel above one milliwatt. Thus, 10 dBm represents a signal 10 dB above or higher than 1 mW, while 20 dBm represents a signal 20 dB above 1 mW, and so on. Since a 30-dBm signal is 30 dB or 1000 times larger than a 1-mW signal, this means that 30 dBm is the same as 1 W. We can use this relationship to construct the watts-to-dBm conversion table that is contained in Table 2.2.

In the wonderful world of transmission over fiber optics, the decibel-milliwatt (dBm) is used to reference optical power to 1 milliwatt. However, in addition to dBm you may periodically encounter the term dBμ, so let's turn our attention to that term.

DECIBEL ABOVE 1 μW The term dBμ is used for optical power referenced to 1 microwatt (μW). Thus,

$$\text{dB}\mu = \frac{\text{output power}}{\text{1-}\mu\text{W input}}$$

Power Budget

Many times when we work with fiber optics we will encounter the term *power budget* without an explanation of its meaning. As we have just discussed decibels we will briefly discuss power budget, although we will return to this metric when we cover fiber in the next chapter.

TABLE 2.2 Relationship between Watts and Decibel-Milliwatts

Power in watts	Power in dBm, dBm
0.1 mW	−10
1 mW	−0
1 W	−30
1 KW	−60

The term power budget is used to reference the difference in dB between the transmitted optical power, typically expressed in dBm, and the receiver sensitivity, also commonly expressed in dBm. dB is used, as the connection of circuits in tandem permits the power levels in decibels to be arithmetically added and subtracted. For example, if a known amount of optical power, in decibel-milliwatts, is inserted or launched into a fiber, and the losses, in decibels, for the various components such as lengths of fiber and connectors are known, the overall link loss can be easily computed with simple addition and subtraction. Because 1 dBμ equals 0.001 dBm, 0 dBm is then equivalent to 30 dBμ. Table 2.3 illustrates the relationship between common values of decibel-milliwatts and decibel-microwatts.

Because we noted the relationship between power in watts and decibel-milliwatts in Table 2.2 and that between decibel-milliwatts and decibel-microwatts in Table 2.3, it becomes possible to plot the relationship of all three metrics. This relationship is illustrated in Figure 2.9.

Signal-to-Noise Ratio

Although someone with a strong optics background may be a bit puzzled by this book describing the signal-to-noise ratio more commonly used to categorize copper- and atmosphere-based transmission systems, there is a reason for this discussion. By understanding the constraints of copper-based transmission systems, we can better understand the similarities and differences concerning the transmission capability over an optical fiber.

TABLE 2.3 Relationship between Decibel-Milliwatts (dBm) and Decibel-Microwatts (dBμ).

dBm	dBμ
0	30
−10	20
−20	10
−30	0
−40	−10
−50	−20
−60	−30

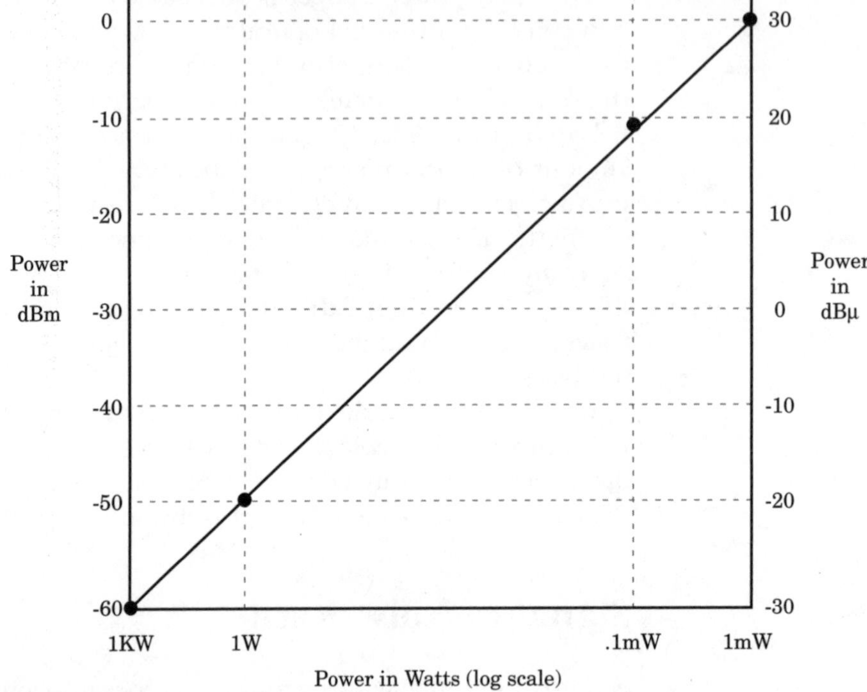

Figure 2.9 The relationship between power in watts, decibel-milli-watts (dBm), and decibel-microwatts (dBμ).

One of the more important metrics in communications is the *signal-to-noise ratio.* In all copper-based communication systems there is a certain degree of noise, caused by the movement of electrons, power-line induction, and cross-modulation from adjacent wire pairs or in wireless communications from frequencies in adjacent channels.

The signal-to-noise ratio is used primarily to categorize transmission on a copper cable; along with bandwidth, it also affects the transmission capacity of a channel. Thus, information about the signal-to-noise ratio is relevant to all types of copper-based transmission media. Two basic categories of noise are relevant for transmission on a copper cable: thermal and impulse noise. *Thermal noise,* such as the movement of electrons or basic radiation from the sun, is characterized by a near-uniform distribution of energy over the frequency spectrum.

Figure 2.10 illustrates thermal noise, which is also referred to as *white* or *gaussian noise.* Because thermal noise is characterized by a near-uniform energy distribution over the frequency spectrum, it represents the lower level of sensitivity of a receiver. This is because a received signal must

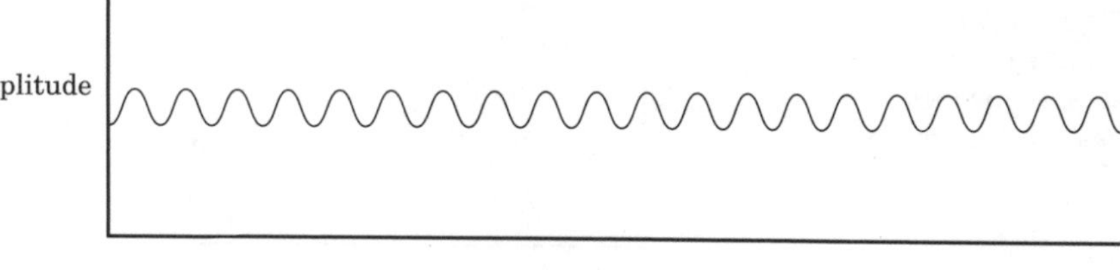

Figure 2.10 White or thermal noise is characterized by a near-uniform distribution of energy over the frequency spectrum.

exceed the level of background noise if it is to be distinguished from the noise emitted by the receiver.

A second type of noise is formed from periodic disturbances, ranging from solar flares, commonly referred to as *sunspots*, to the effect of lightning and the operation of machinery. This type of noise is referred to as *impulse noise* (see Figure 2.11). In Figure 2.11, note that impulse noise consists of irregular spikes of pulses of relatively high amplitude and short duration. As pointed out in Chapter 1, one key advantage of an optical transmission system is its immunity to electromagnetic interference, including solar flares.

The *signal-to-noise ratio* (S/N), which is measured in decibels and is defined as the ratio of the signal power S divided by the noise power N on a transmission medium, is used to calculate the quality of transmission on a copper circuit. Although an S/N ratio greater than unity is always preferred because the receiver must be able to discriminate the signal from the noise, there are limits to how high the S/N ratio can be. These limits are associated with different systems that cap the allowable amount of signal power. These limits may be imposed by system operators or by the FCC (Federal Communications Commission) or another regulatory body; limits established by the latter are designed to minimize interference between different wireless systems.

Table 2.4 provides a summary of the relationship between decibel values and power or S/N ratios. Several of the values listed in Table 2.4 are both significant and warrant some degree of elaboration. First, consider a value of 0 dB. Since the decibel is defined as $10 \log_{10} (P_O/P_I)$, this means that for a decibel reading of zero, $10 \log_{10} (P_O/P_I)$ must be zero. This is possible only if P_O/P_I = unity, which means that a value of 0 dB occurs when the input power equals the output power. We can also note that a 0 dB value means that there is no gain or loss at the termination point of a transmission system.

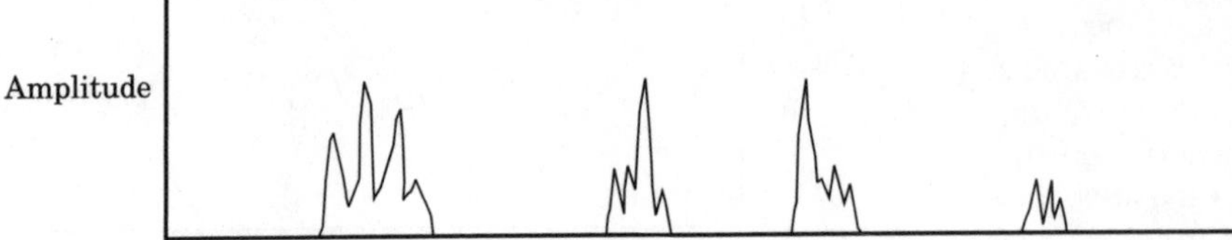

Figure 2.11 Impulse noise occurs at random times at random frequencies.

TABLE 2.4 Relationship between Decibels and Power Measurements

dB	Power or S/N Ratio
0	1.0 : 1
1	1.2 : 1
2	1.6 : 1
3	2.0 : 1
4	2.5 : 1
5	3.2 : 1
6	4.0 : 1
7	5.0 : 1
8	6.4 : 1
9	8.0 : 1
10	10.0 : 1
13	20.0 : 1
16	40.0 : 1
19	80.0 : 1
20	100.0 : 1
23	200.0 : 1
26	400.0 : 1
29	800.0 : 1
30	1000.0 : 1
33	2000.0 : 1

TABLE 2.4
(Continued)

dB	Power or S/N Ratio
36	4000.0 : 1
39	8000.0 : 1
40	10000.0 : 1
50	100000.0 : 1

A second important value is 3 dB, which is equivalent to a power or S/N ratio of 2:1. Thus, a 3 dB value indicates that the signal value is twice that of the noise. Finally, if you scan Table 2.1 and focus on decibel values in increments of 10, you will note that they correspond to S/N ratios that increase by a power of 10. Thus, a decibel value of 10 is equivalent to an S/N ratio of 10, a decibel value of 20 is equivalent to an S/N ratio of 100, and so on.

Metallic Media Transmission Rate Constraints

In a copper cable or atmospheric transmission system two key constraints govern the ability to transmit information at different data transmission rates: the Nyquist relationship and Shannon's law.

Nyquist Relationship

The *Nyquist relationship* governs the signaling capability on a channel. In 1928, Harry Nyquist developed the relationship between the bandwidth and the signaling rate (baud) on a channel as $B = 2W$ where B is the baud rate and W is the bandwidth in hertz.

BITS VERSUS BAUD Before discussing what the Nyquist relationship means in terms of the maximum achievable signaling rate on a channel, a brief digression into bits and baud rate is warranted. The bit rate, typically presented in terms of bits (of data transmitted) per second (bits/s), represents a measurement of data throughput. In comparison, the baud represents the rate of signal change commonly expressed in terms of hertz. When information in the form of bits is to be transmitted, an oscillating wave is varied to impress or modulate information. The oscillating wave is referred to as a *carrier.* Common modulation techniques include altering the amplitude of the carrier [amplitude modulation (AM)], altering the frequency of the carrier [frequency modulation (FM)], and altering the time period or phase of the carrier [phase modulation (PM)]. Some communication systems also alter two characteristics of the carrier, such as amplitude and phase.

We can commence our probe into the relationship of bits and baud with a simple modulation scheme referred to as *frequency-shift keying* (FSK), where each bit is modulated using one of two tones that we can refer to as f_1 and f_2. If we assume that all binary 1s are modulated at f_1 and all binary 0s with f_2, or vice versa, this simple modulation scheme results in each bit being equivalent to one signal change. Thus, in this situation the bit rate equals the baud rate.

Now let's apply a more sophisticated modulation technique. Under phase modulation, we can vary the phase of a signal according to the composition of a single bit or a group of bits. Suppose our phase modulation technique involves varying the phase of the carrier to one of four positions (0°, 90°, 180°, and 270°). This modulation technique would then allow each possible combination of 2 bits to be encoded into one signal change. An example of this technique is shown in Table 2.5.

In examining the relationship between bit pairs and phase change, it is clear that the bit rate is twice the signaling rate since 2 bits are packed into each signal change. The technique described above is referred to as *dibit encoding* and represents one of many types of data modulation techniques. For now, the important aspect of this digression is to note that the bit rate may or may not equal the baud rate, depending on the method of data encoding used by a modulation scheme. Now that we have an appreciation for the difference between bit and baud rates as well as the need to modulate both data and voice for wireline and wireless transmission, let's end this digression and return to our discussion of the Nyquist relationship.

MAXIMUM MODULATION RATE The *Nyquist relationship* states that the maximum rate at which data can be transmitted before one symbol interferes with another, a condition referred to as *intersymbol interference,* must be less than or equal to twice the bandwidth in hertz.

TABLE 2.5 Potential Mapping of Bit Pairs into Phase Changes

Bit pair	Phase change
00	0
01	90
10	180
11	270

Because most transmission systems modulate or vary a signal, the Nyquist relationship limits the signaling rate, which is proportional to the available bandwidth. This explains why modem designers use different techniques to pack more bits into each baud to achieve a higher data transfer rate. Because the bandwidth of a telephone channel is fixed, a modem designer cannot exceed a given signaling rate before intersymbol interference adversely affects the data flow. Thus, to transmit more data, the modem designer must pack more bits into each signal change. In the telephone company infrastructure a bandwidth of 4 kHz is provided for a voice channel, which explains why the pulse-code modulation (PCM) sampling rate is 2×4 kHz or 8000 times per second. In addition to the Nyquist relationship, which governs the signaling rate, there is a second constraint that limits the maximum achievable data rate obtainable on a copper transmission medium. That constraint is Shannon's law.

SHANNON'S LAW In 1948, Professor Claude E. Shannon presented a paper concerning the relationship between coding and noise and computed the theoretical maximum bit rate capacity of a channel of bandwidth W in hertz as follows:

$$C = W \log_2 \left(1 + \frac{S}{N}\right)$$

where C = capacity in bits per second, bits/s
W = bandwidth, Hz
S = power of the transmitter
N = power of thermal noise

In 1948, a "perfect" telephone channel was considered to have an S/N ratio of 30 dB, which represents a value of 1000. The maximum data transmission for a telephone channel in 1948 was then proposed as

$$C = W \log_2 (1 + \frac{S}{N})$$

$$W = 3000 \log_2 (1 + 10^3)$$

$$S = 3000 \log_2 (1001)$$

$$N \cong 3000 \text{ bits/s}$$

It should be noted that now, so many years after Shannon's paper was published, the maximum data transmission rate on an analog telephone channel is still 33.6 kbits/s, which is within 10 percent of Shannon's law. In reality, telephone channels today have a bit (no pun intended) less noise. Although you are probably well aware of 56-kbit/s modems, they operate near that rate only downstream where the destination location has a direct digital connection. In the upstream direction the 56-kbit/s modem is still limited to a data transmission rate of 33.6 kbits/s.

Optical Media Transmission Rate Constraints

In an optical transmission environment information can be transmitted by altering the light source between ON and OFF states. Three key factors affect the transmission of light pulses on an optical fiber: attenuation, dispersion, and fiber nonlinearities resulting from the fact that no fiber is pure glass.

Attenuation

Attenuation in an optical fiber is closely similar to attenuation of a signal transmitted on a copper circuit. In an optical environment attenuation represents the diminution of optical power, where light pulses become smaller and distorted as light flows down a fiber. Attenuation results primarily from absorption and scattering; scattering is a function of fiber nonlinearities and chromatic dispersion.

Optical attenuation is usually expressed in decibels without a prefix or a negative (minus) sign even though optical power is lost. However, in this book we will use a negative-sign prefix to indicate any calculations that show a loss. Another closely related term that warrants attention is the *attenuation coefficient,* which represents the loss of optical power per a given length of fiber, typically per kilometer. Thus, the attenuation coefficient is expressed in terms of dB/km.

Scattering

The purity of an optical fiber also governs the obtainable data (transmission) rate, although it does so indirectly, which requires a bit of explanation. As light flows through a fiber and encounters an impurity, the direction of flow of the photon hitting the impurity will change. This results in an effect referred to as *scattering,* which spreads the pulse and makes it more difficult to detect at the receiver.

Chromatic Dispersion

Another factor that limits the pulse rate of optical transmission is *chromatic dispersion,* which results from the fact that the speed of an optical pulse traveling in a fiber changes as the pulse's wavelength changes. As photons flow through a fiber, the 2- to 5-nm spectral width of the laser results in a broadening of pulses, especially as photons scatter. The net effect of attenuation, scattering, and chromatic dispersion is illustrated in Figure 2.12. The top portion of the figure illustrates a sequence of "perfect" pulses generated by a laser. The middle sequence illustrates the effect of attenuation on the pulses, while the lower portion of the illustration indicates the effect of scattering and chromatic dispersion.

In examining Figure 2.12, note that as the transmission distance increases, the pulses become smaller as attenuation increases and wider as the effect of scattering and chromatic dispersion increases.

Operating Rates

Currently, practical optical transmission systems operate at 10 Gbits/s. However, optical transmission at 40 Gbits/s is expected to become commercially available in the near future. For data rates of up to 2.5 Gbits/s,

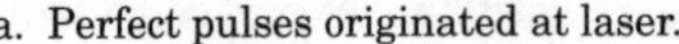

Figure 2.12
The combined effect of attenuation, scattering, and chromatic dispersion on the flow of light pulses down a fiber.

a. Perfect pulses originated at laser.

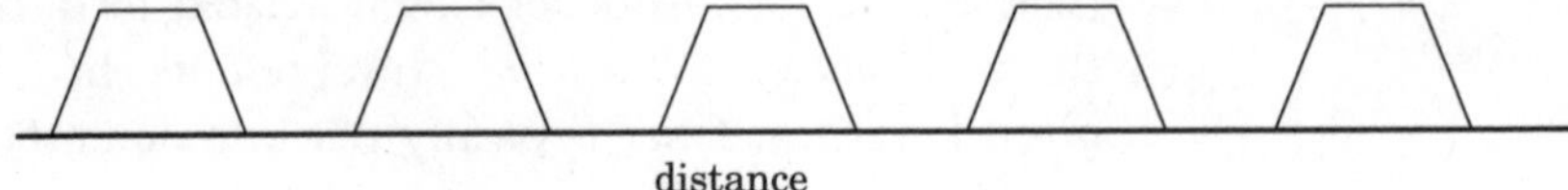

b. Effect of attenuation.

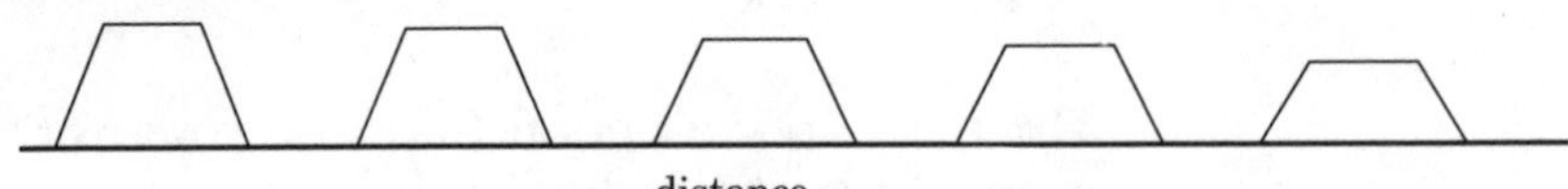

c. Effect of scattering and chromatic dispersion and attenuation.

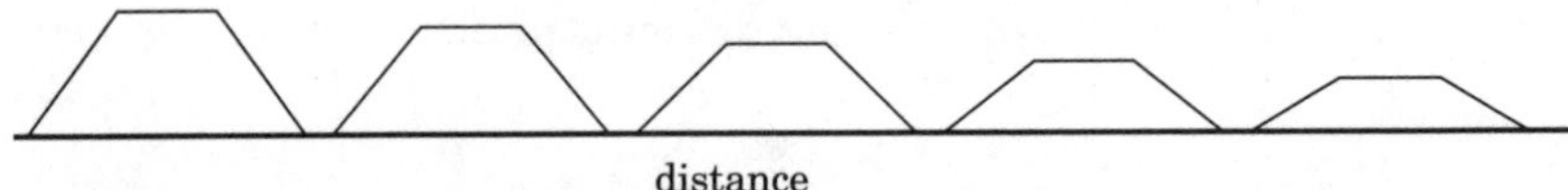

most optical transmission systems use lasers that are switched on and off, representing digital 1s and 0s. At higher data rates a modulation containing a crystal lithium niobate waveguide is commonly used with a laser. The laser beam is fired into the modulator, which splits the beam into two parallel beams, each with half the amplitude of the original. One of the beams passes through a slice of lithium niobate, which, when a voltage is applied to the crystal, refracts the light and causes the beam to become out fo phase with the other beam. When the beams are joined together at the other end of the modulator, they cancel each other out, resulting in an OFF condition or 0 data bit. When the voltage is removed from the lithium niobate, the beams remain in phase with each other and combine to produce a beam with the same amplitude as in the original beam. This signifies a "1" data bit. Because it is possible to apply or remove a voltage to or from a lithium niobate crystal faster than switching a laser on and off, this type of modulator offers the potential to break the present speed barrier for transmitting over optical fiber.

Distance also limits the speed of transmission over an optical fiber. Because transmission distance is a function of the optical transmitter and type of optical fiber, we will postpone a discussion of the constraints associated with transmission distance to Chapters 3 and 4.

CHAPTER 3

Understanding Optical Fiber

The goal of this chapter is to explain the production, operation, and utilization of optical fiber. In so doing we will describe how light travels down an optical fiber, discuss the different types of fiber available for different applications, and examine several parameters that govern the ability to transmit information in the form of light efficiently and effectively through an optical fiber.

Evolution

Although we associate the transmission of light within an optical fiber as a modern element of science, the concepts behind the technology date to the nineteenth century. During the mid-1800s the physicist John Tyndall showed that light could be bent around a corner while it traveled through a stream of pouring water. During 1880, Alexander Graham Bell, who we associate primarily with the invention of the telephone, demonstrated the use of a membrane to modulate an optical signal in response to varying sound. Bell's photophone represented a free-space transmission system and not a guided optical system; however, it paved the way for further effort.

Although AT&T obtained a patent on guided optical communications over glass in 1934, at that time the glass manufacturing process did not provide the capability needed to produce fiber-optic cable with an attenuation level low enough to make guided optical communications a reality. Instead, approximately 30 more years passed until researchers were able to better understand how light attenuates in glass and how optical fiber should be manufactured to provide a practical method for supporting optical communications. The efforts of various researchers resulted in a reduction in the attenuation of glass-fiber optic cable from over 1000 dB/km to under 20 dB/km. In 1970, Corning Glass Works patented its fabrication process, which made it possible to manufacture fiber with a loss of 20 dB/km. That level of loss was originally considered as a Rosetta stone for optical communications. This is because a loss of 20 dB/km is equivalent to receiving 1 percent of the original light power after traveling a distance of 1 km. Today you can obtain optical fiber whose attenuation can range to below 0.5 dB/km, illustrating the progress that has occurred in the manufacture of optical fiber since the late 1960s or the early 1970s. Of course, as you might expect, the attenuation of an optical fiber governs the fiber's production cost and the resulting retail price. Companies today, ranging in scope from small-business LAN operators to multinational telecommunications giants, can select from a wide range of optical fibers with attenuation limits suitable for different applications.

Fabrication

Two basic methods are used to manufacture optical fiber. One method requires the deposit of chemical vapor or gas with impurities within a tube prior and during its heating. The soot produced by the heating process results in impurities that define the index of refraction (refractive index). As we will note later in this chapter, the refractive index difference between the core and the outer area of a fiber, which is referred to as the *cladding*, results in light pulses remaining inside the fiber. The deposit of chemical vapor or gas is referred to as *modified chemical vapor deposition* (MCVD). A second fiber fabrication process results in vapor deposits placed on the outside of the tube. This process is referred to as *outside vapor deposition* (OVD).

Traversing Burning

Both MCVD and OVD reference the location of the vapor with respect to the glass tube. The actual fabrication process that uses the glass tube to create a spooled wire employs a technique referred to as *traversing burning*. This technique consists in rotating a large glass tube over a heat source while passing a series of gases through the tube. The soot from the gases is deposited either inside (MCVD) or outside (OVD) the tube until a certain level is reached where the flow of gases is stopped. At that time the heat is increased and the sooted glass tube is stretched. The resulting solid rod of glass is then placed in a furnace where it is melted and allowed to trickle down through an opening to form a fiber-optic strand. As the fiber exits the furnace, it is coated and spooled by a winder. The coating of the fiber results in an epoxy or plastic film added to the glass to provide protection. This process, which is referred to as a *fiber-drawing tower*, is illustrated in Figure 3.1.

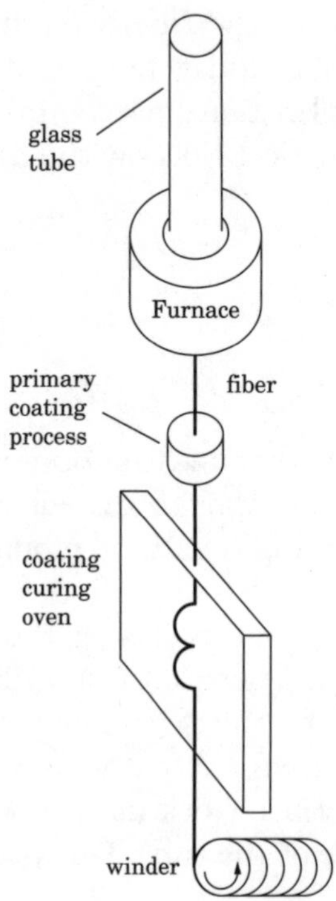

Figure 3.1
A popular method of manufacturing fiber strands employs a fiber-drawing tower through which a glass compound is heated, stretched, and spooled.

The process illustrated in Figure 3.1 represents a general schematic of a fiber-drawing tower. Not shown but of extreme importance are feedback control systems ensuring that the diameter and the level of impurities of the fiber govern its index of refraction within predefined thresholds. Once the fiber is spooled, it will be used as input to another manufacturing process that adds a strength member and jacket to form an optical fiber cable.

Although the first type of optical fiber was glass-based, during the late 1980s and 1990s other types of optical fiber were produced using plastic. Similarly, the fabrication of a single-fiber cable resulted in the development of a variety of cable types, which we will examine later in this chapter.

Basic Composition

Once a spool of optical fiber is manufactured, it has to be fabricated into a cable. The spooled fiber represents the core of a fiber-optic cable, and the coating is referred to as the *cladding*.

Cladding

The *cladding* consists of glass or plastic of a density different from that of the core and functions as a mechanism to contain the light signal within the fiber. Both the core and the cladding are manufactured together during the MCVD or OVD process described above. As we will soon note, the core and cladding have different indices of refraction that determine how light is guided down the core of the cable.

Jacket

During a second manufacturing phase, an additional coating, referred to as the "jacket," is applied around the cladding. Representing the outer layer of the cable, the jacket is usually colored orange, although this author has seen yellow and black fiber-optic jackets.

The jacket usually consists of one or more layers of polymer that insulate the core and cladding from shock as well as damage from the outside environment, such as water, solvents, and abrasions. Note that the jacket lacks any optical properties and does not affect the propagation of light within the cable.

Strengthening Fibers

Most types of fiber-optic cables include strengthening fibers that help protect the relatively thin core from crushing forces and excessive tension during the cable installation process. The material used as strengthening fibers can range in scope from wire strands to Kevlar 4 to gel-filled sleeves. The top portion of Figure 3.2 illustrates a three-dimensional side view of a basic fiber-optic cable; the lower portion shows the cable with respect to a cross-sectional view.

As progress was made in manufacturing fiber-optic cable, the basic cable design underwent several modifications. One popular modification was the use of a concentric buffer area to shield the core and cladding from damage; another modification involved the use of a concentric strength material instead of a wire. Figure 3.3 is a cross-sectional view of this revision to the basic fiber-optic cable.

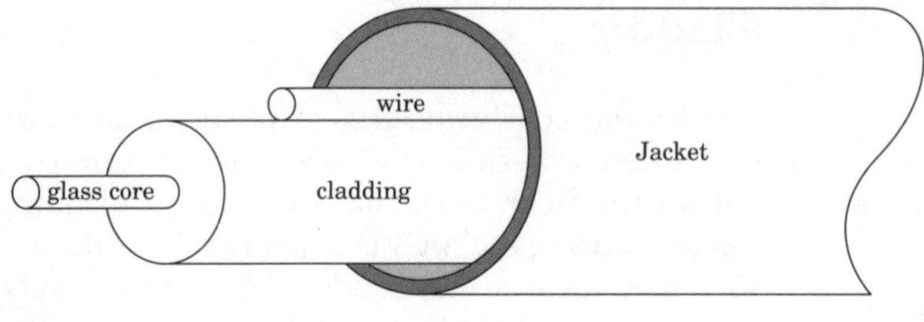

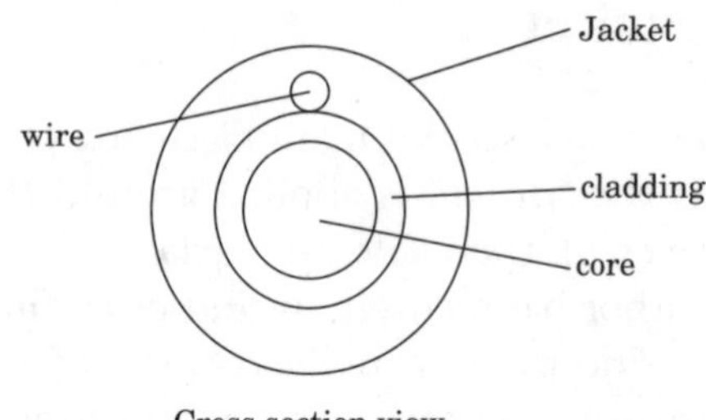

Figure 3.2 The components of a basic fiber-optic cable.

Light Flow in a Fiber

The ability of light to flow down the core of a fiber-optic cable results from the fact that the core and cladding have different indices of refraction. Let's discuss this topic.

Index of Refraction

The *index of refraction* of a material is the ratio of the velocity of light in a vacuum to that in a material. Thus, the index of refraction, commonly denoted in physics books as n, can be defined as $n = c/v$, where c is the speed of light in a vacuum and v represents the speed of light in a medium. Since v is always less than c, the index of refraction is a dimensionless number greater than one.

Figure 3.4 illustrates the bending of light as it passes between two or more media that have different indices of refraction. In this example we will assume that the beam of light has a wavelength of 550 nm traveling while in air and arrives at the junction with the second material at an angle of 40.0° (01) with the normal. The refracted beam is shown, forming an angle of 26.0° (02) with the normal. Without further information, it is possible to compute the index of refraction using Snell's law of refraction:

$$n_1 \sin 0_1 = n_2 \sin 0_2$$

where n_1 and n_2 are the indices of refraction for each material and 0_1 and 0_2 represent the incident-beam angle with the normal and the refracted beam angle with the normal, respectively.

A second factor that enables light to propagate down the length of an optical fiber is based on the angle of the rays injected into the fiber. For light rays to be reflected, the rays must be injected at incident-beam angles less than what is referred to as the *critical angle*. Thus, before proceeding further, let's again digress and examine the role of the critical angle.

The Critical Angle

The *critical angle* is critical as it determines why your digitized phone calls, Internet surfing efforts, and other information flowing as light within an optical fiber doesn't leak out of the fiber. Thus, when a beam of light hits the juncture between the core and the cladding, the light beam will be totally refracted back into the core if the angle of the beam with respect to the juncture is greater than some specific value. That value is the critical angle, and this phenomenon is referred to as *total internal reflection* (TIR); it is called "total" as virtually 100 percent of the beam is reflected.

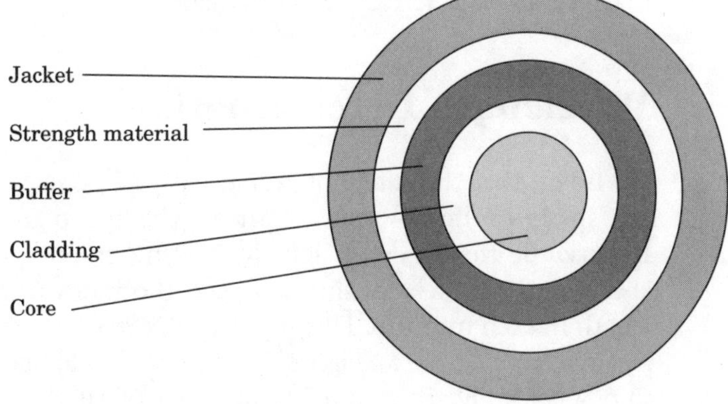

Figure 3.3
A cross-sectional view of a popular revision to the basic fiber-optic cable.

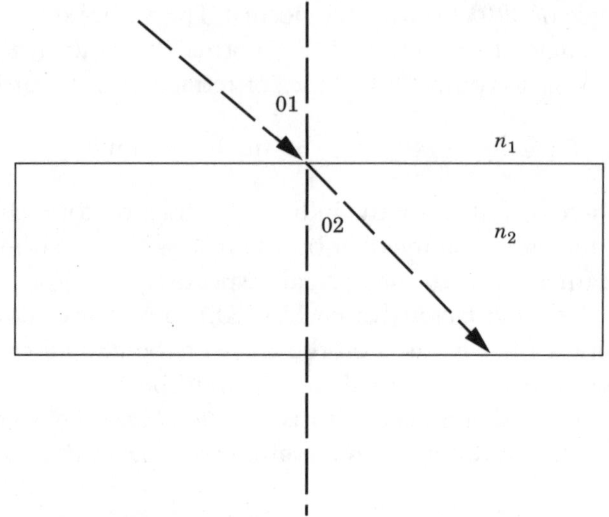

Figure 3.4
The refraction of a beam of light is based on the indices of refraction (n_1 and n_2) for each material. According to Snell's law of refraction, $n_1 \sin 0_1 = n_2 \sin 0_2$.

Figure 3.5 illustrates the flow of light with respect to the critical angle. Note that the farther the beam is from the perpendicular when it strikes the juncture, the more strongly it is bent. We noted from Snell's law that $n_1 \sin 0_1 = n_2 \sin 0_2$. Thus, we can use Snell's law to express the critical angle as follows:

$$\cos 0_c = \frac{n_{cladding}}{n_{core}}$$

where 0_c is measured from the axis of the core of the fiber. To obtain total internal refraction, the index of refraction of the core (n_{core}) must be greater than the index of refraction of the cladding = $n_{cladding}$ or $n_{core} > n_{cladding}$.

Wavelength Determination

We noted that the index of refraction can be expressed by the ratio of the speed of light in a vacuum to the speed of light in a medium. Because the frequency of light is the same in any given medium, we can also express the relationship between the index of refraction and wavelength in each medium. Thus, we can express that relationship as follows for two media: $n_1\lambda_1 = n_2\lambda_2$. Figure 3.4 illustrates an application of this relationship. Specifically, air has a refractive index of 1.0; thus, $n_1 = 1.0$. Let's assume

Figure 3.5
The critical angle.

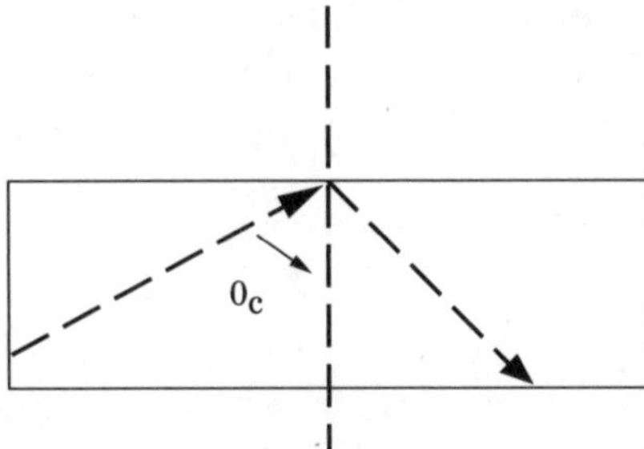

When the angle is greater than 0c light is totally reflected back into the core.

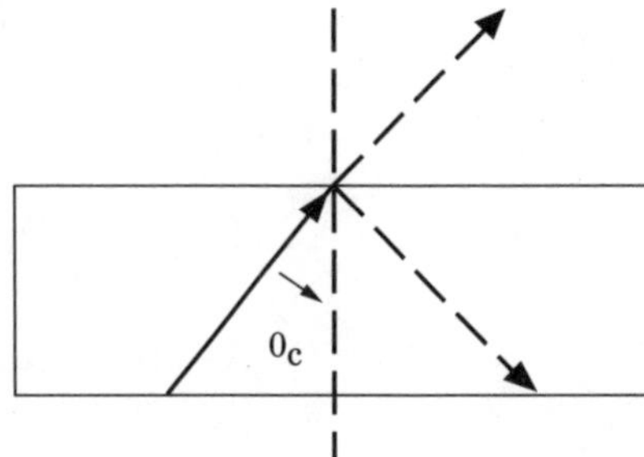

When the angle is less than 0c some light leaves the core.

that the second medium has an index of refraction of 1.47; thus, $n_2 = 1.47$. Because we previously assumed that the initial beam of light had a wavelength of 550 nm, we can easily compute the wavelength in the second medium. Because $n_1\lambda_1 = n_2\lambda_2$, we obtain 1.0×550 nm $= 1.43 \times \lambda_2$. Thus

$$\lambda_2 = \frac{550\ nm}{1.43} = 374\ nm$$

In addition to the participation of the refractive index, wavelength, and critical angle in the flow of light in a fiber, another term you will frequently encounter is *numerical aperture*. Thus, prior to discussing the flow of light through an optical fiber, let's focus on what the numerical aperture is and how it governs transmission via an optical fiber.

Numerical Aperture

The *numerical aperture* (NA) of an optical fiber provides a measurement of the light-gathering capability of a fiber. Mathematically, the NA reflects the difference in the refractive index of the core and cladding as follows:

$$\text{NA} = \sqrt{n^2_{\text{core}} - n^2_{\text{cladding}}}$$

where n_{core} represents the index of refraction of the core of the fiber while n_{cladding} represents the index of refraction of the cladding that surrounds the core.

As the value of NA increases, the optical fiber permits more light to be transmitted through the fiber. Now that we have seen the relationships associated with the indices of refraction of the core and cladding of a fiber, let's determine how light flows within an optical fiber.

Light Flow

As discussed previously, the ability to ensure that light flows or propagates the length of an optical fiber is based on ensuring that the ray is totally refracted whenever it reaches the core-cladding boundary. This is indicated by the numeral 1 in Figure 3.6. To accomplish this, light rays must be injected into the fiber at an angle less than the critical angle, resulting in total internal reflection (TIR).

Figure 3.6 illustrates the potential flow of a light ray within an optical fiber. As noted by the numerals 1, 2, and 3, three potential scenarios can govern a light ray's flow. First, the ray can be injected and focused directly into the core. This requires the use of a laser as the light source and a special type of optical cable referred to as a *single-mode cable*. The term *single mode* reflects the fact that there is no refraction and the pulses that are injected into the fiber travel in one mode. As we will note later in this chapter, the ability to accomplish single-mode light propagation requires the use of a relatively small core, which makes the fiber more expensive to manufacture.

A second method of light flow results in the reflection of light at the core-cladding boundary. As indicated by the numeral 2 in Figure 3.6, the rays are reflected when the angle that the ray forms with the boundary is less than the critical angle. This results in total internal reflection.

Because the use of light-emitting diodes (LEDs) results in a spread of rays entering an optical fiber, the result is a series of reflected rays that are received at the opposite end of the fiber that "spreads" the original injected light pulse. This situation, referred to as *modal dispersion,* requires the transmitter to slow its light pulse rate to a pace slower than that of the dispersion at the receiver to allow pulses to be recognized. Thus, the bouncing effect also affects the data transfer rate. As we will discuss later in this chapter, two types of optical fiber allow multiple modes to flow: step-index and graded-index multimode optical fiber.

Returning to Figure 3.6, in a third light flow scenario, the angle of the light ray is greater than the critical angle. This is shown by the numeral 3 in Figure 3.6. When this situation occurs, a portion of the ray will be par-

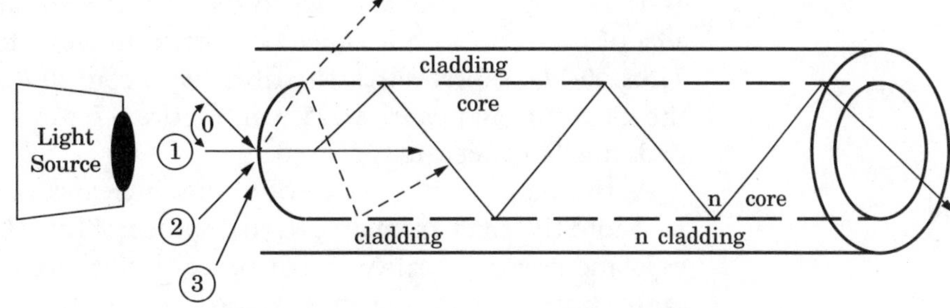

Figure 3.6
The flow of light rays into and through an optical fiber.

The ability to transmit light through an optical fiber depends upon the focusing capability of the light source and the indices of refraction of the core and cladding. In (1) a laser is used with single mode fiber to direct a ray through the fiber without any refraction. In (2) the angle of a ray is less than the critical angle, resulting in the ray bouncing along the core/cladding boundary. In (3) the angle exceeds the critical angle, resulting in a portion of the ray refracted into the cladding and a portion refracted into the core.

tially refracted into the cladding, where it will eventually be lost while a portion will be reflected back into the core.

Now that we understand the basic principles of fabrication and general flow of light within an optical fiber, let's try to define the characteristics of this light conductor or waveguide. First, let's discuss cable metrics and terms that define different optical fiber characteristics.

Optical Fiber Metrics and Terms

If you study the specification sheet (specs) produced by a manufacturer of fiber-optic cable, you will note a series of terms that specify various aspects of the cable. In this section we discuss those terms and attempt to understand their meaning.

Cable Size

In one common method of specifying a fiber-optic cable, size of the cable's core and cladding areas is a key factor. If you use this method, you will usually encounter a pair of numbers separated by a slash, such as 50/125. These numbers refer to the diameter of the core and cladding area of the cable in micrometers, where a micrometer represents 10^{-6} m. Thus, 50/125 indicates that the fiber has a core 50 μm in diameter while the cladding diameter is 125 μm. Figure 3.7 shows how the core and cladding diameters are specified.

As the diameter of the core of a fiber increases, more light can be coupled into the fiber from an external source. While this may appear to be welcome news, in reality it can be equivalent to a child let loose in a candy store since too much of a good thing can be bad. In other words, too much light can result in saturation of a receiver. Thus, there are constraints on the size of the core. Because one of those constraints is based on the general category of the optical fiber, let's discuss this topic.

Cable Category

There are two general categories of optical fiber: single-mode and multimode. Because multimode fiber was commercialized prior to single-mode, let's review their characteristics in their order of commercialization.

MULTIMODE FIBER *Multimode fiber* has a much larger core than single-mode fiber; therefore, multimode fiber permits hundreds to thousands of rays or modes of light to simultaneously propagate through the fiber. This type of fiber permits light to enter from different angles; as a result, the light bounces along the inside of the fiber, reflected back into the core when it hits the core-cladding boundary. This repeated bouncing continues until the light reaches the end of the fiber.

The top portion of Figure 3.8 illustrates this light bouncing effect, which results in pulse spreading or modal dispersion. Because modal dis-

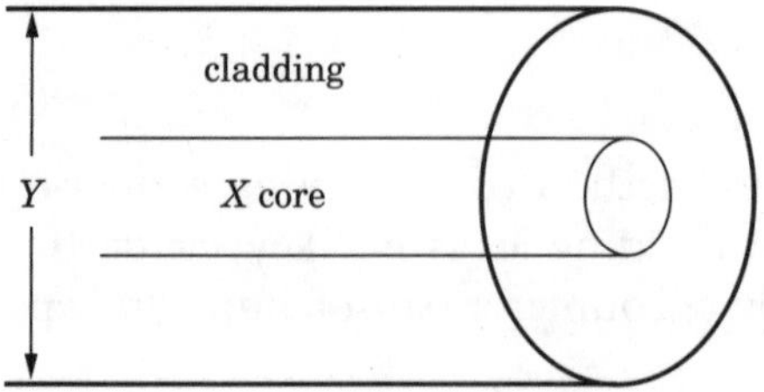

Figure 3.7 The size of a fiber-optic cable is specified in terms of the cable's core *X* and cladding *Y* diameters as *X/Y* in micrometers.

persion is caused by the difference in the propagation times of light rays, which take different paths along the fiber, a stretched light pulse at the end of the fiber results. That stretched light pulse is reduced in amplitude at the output of the fiber, making it more difficult for the receiver to detect. Thus, in general, multimode fiber is used for the transmission of light over relatively short distances, such as in a LAN environment.

STEP-INDEX FIBER The flow of light shown at the top of Figure 3.8 represents the use of one of two types of multimode fiber. In this example the flow of light is shown with respect to the use of a step-index multimode fiber. This type of optical fiber has a large core, usually up to 100 μm in diameter. As shown in Figure 3.8, some of the rays that make up a light pulse can travel a direct route from end to end, which is denoted as a direct ray. In contrast, other rays bounce off the cladding and arrive at different times at the end of the fiber. Because this action spreads the pulse, another problem resulting from the use of step-index multimode fiber is that the pulses must be spaced sufficiently far apart at the transmitter to prevent them from overlapping at the receiver, enabling the receiver to discriminate between pulses. This need for a time delay between pulses limits the pulse rate, which, in turn, limits the amount of information per unit of time that can be transmitted. Although this

Figure 3.8 Comparison between light flow in multimode step-index and graded fibers.

a. Multimode step-index fiber.

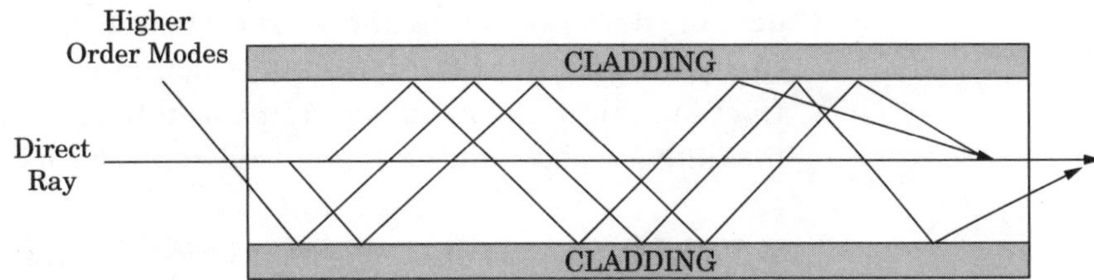

b. Multimode graded-index fiber.

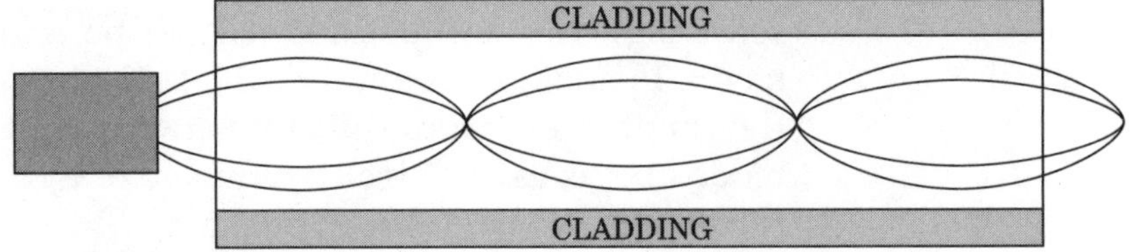

is a constraint, multimode fiber is suitable for use with an LED light source, which is less costly than a laser.

GRADED-INDEX FIBER A second type of multimode fiber contains a core in which the refractive index gradually diminishes from the center of the core outward toward the cladding. Thus, this type of fiber can be considered to have a variable refractive index. The higher refractive index at the center of the core results in the transmission of light rays along the axis occurring more slowly than the transmission of rays near the cladding. In addition, because of the graded index, light in the core follows a helix-shaped curve instead of bouncing off the cladding. The shortened path and higher speed obtainable because of the lower index toward the cladding results in the series of rays arriving at the receiver at approximately the same time.

The lower portion of Figure 3.8 illustrates the flow of light through a multimode graded-index fiber. This type of fiber is more costly to manufacture than a step-index fiber. Popular graded-index fiber-optic cables have core diameters of 50, 62.5, and 85 μm, with a cladding diameter of 125 μm. The 62.5/125 fiber-optic cable is one of the most popular fibers used in data communications applications. Because a multimode graded-index fiber minimizes modal dispersion, it supports a higher data transmission rate than does a multimode step-index fiber.

Although multimode fiber is used primarily in LAN applications, this was not always the case. As communications carriers originally developed an optical backbone in the late 1970s or early 1980s, they first turned to the use of multimode fiber. However, because single-mode fiber supports much higher transmission rates, it is now the preferred medium for backbone and long-haul applications.

SINGLE-MODE FIBER As mentioned above, a second general category of fiber is single-mode fiber. This type of fiber has a much smaller core diameter and allows only one mode of light at a time to propagate through the core; for this reason, there is no modal dispersion or pulse spreading. Thus, an optical signal can be transmitted over longer distances. In addition, because of the lack of pulse spreading, the optical transmitter can reduce the gaps between pulses necessary when multimode fiber is used. This allows the use of a faster light source, which in turn supports a higher operating rate.

Figure 3.9 illustrates the transmission of a light ray through a single-mode fiber. The core of a single-mode fiber is typically 8 to 10 μm, which means that it is more expensive to manufacture than multimode fiber. The outer cladding diameter of most single-mode fibers is 125 μm. Thus, common size specifications will indicate 8/125 or 10/125 for the size of single-mode fiber.

The index of refraction between the core and the cladding for single-mode fiber is near uniform, resulting in light traveling parallel to the core axis. The primary use of single-mode fiber is for long-distance communications, with millions of kilometers of this type of fiber installed each year.

Optical Attenuation

In communications we use the term *attenuation* to denote a loss of signal power. In the wonderful world of optical communications, we use the rate of optical power loss with respect to distance along the fiber to denote attenuation. The loss is usually measured in decibels per kilometer (dB/km) at a specific wavelength.

Attenuation of an optical signal has many causes. As photons flow down an optical fiber, they can be absorbed by collisions with impurities in the fiber, or the photons may scatter, reducing their energy level in electronvolts. Because each wavelength is transmitted differently within an optical fiber, attenuation must be measured at specific wavelengths.

The amount of optical attenuation due to absorption and scattering of light at a specified wavelength (λ) is expressed as follows:

$$A\,(\text{dB}) = -10 \log \frac{P_I}{P_O}$$

where P_I represents the input power and P_O represents the output power.

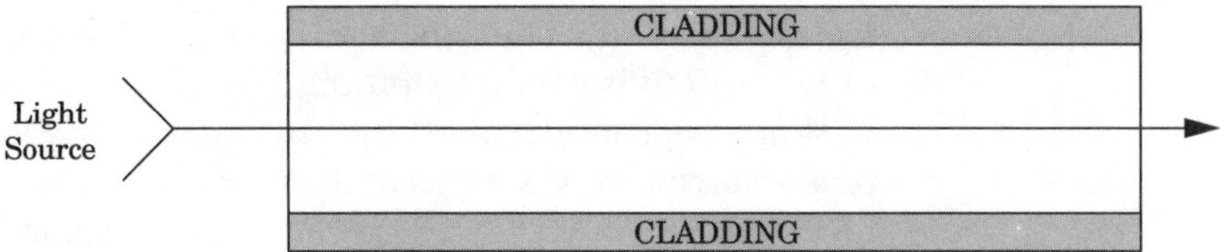

Figure 3.9 The flow of light via a single-mode fiber follows the center of the core.

ATTENUATION AND WAVELENGTH In high school physics classes, instructors often talk about relationships between physical parameters. In the world of electrical communications, we probably heard the expression "high frequencies attenuate more rapidly than low frequencies." If we put on our thinking caps and remember that wavelength is inversely proportional to frequency, we can express a general relationship between attenuation and wavelength. That general relationship can be expressed as "attenuation decreases as the wavelength increases." In more technical terms, the dispersion of electromagnetic radiation varies inversely to the fourth power of the wavelength. This dispersion or scattering is referred to as *Rayleigh scattering*, after Lord Rayleigh, who in 1871 published a paper that described this phenomenon. Rayleigh scattering explains the blue color of the sunlit sky, since blue light, which is located at the short-wavelength portion of the visible spectrum, will be scattered more strongly than longer wavelengths, such as green, yellow, orange, and red. Thus, when we view the sky in a direction other than toward the sun, we observe scattered light, which appears blue. Mathematically we can denote the Rayleigh effect as $I = 1/\lambda^4$.

As an example of the Rayleigh effect, scattering at 400 nm becomes 9.4 times as great as that 700 nm when a person views a common location. Thus, if we plot attenuation as a function of wavelength and consider the effect of Rayleigh scattering only, the relationship would appear as shown in the top portion of Figure 3.10. If we consider only the Rayleigh effect, we would always want to operate at a relatively long wavelength to minimize attenuation. However, another important characteristic of light that we need to consider is its absorption. Absorption is caused by residual impurities in the fiber. In addition, it is nonlinear, with absorption higher at some wavelengths than others. This is shown in the middle portion of Figure 3.10.

In case you are wondering about the bump in the degree of attenuation beginning at 1350 nm, that increase results from the effect of light between 1350 and 1450 nm in the presence of hydroxyl radicals in the cable material. In some publications you will note the term *OH absorption* at the top of the peak at approximately 1400 nm to denote the effect of hydroxyl radicals. Thus, the actual attenuation spectrum as a function of wavelength is a complex process.

When we consider both the Rayleigh effect and absorption, the plot of attenuation versus wavelength appears as shown in the lower portion of Figure 3.10. If you focus on the vertical bars in the lower portion of

Figure 3.10
Optical fiber attenuation as a function of wavelength.

a. Rayleigh scattering results in a wavelength dependence of $1/\lambda^4$.

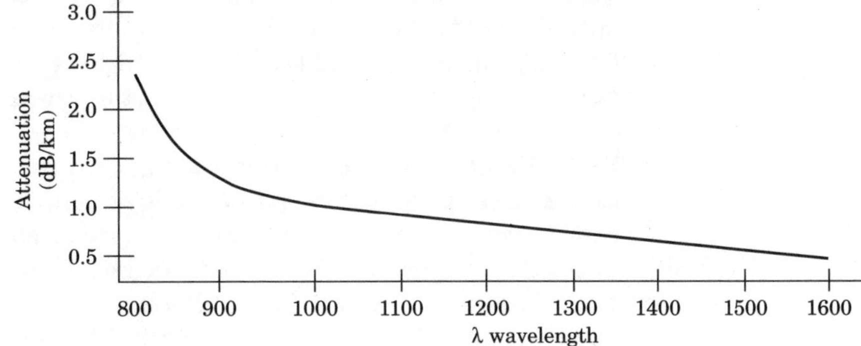

b. Light absorption.

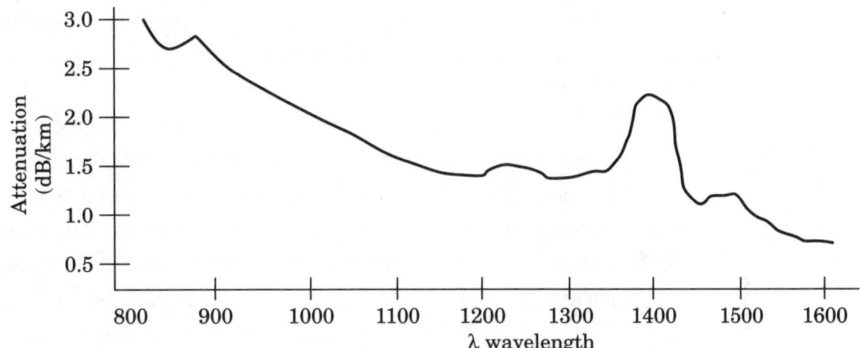

c. Total attenuation from Rayleigh scattering and 1600 nm light absorption.

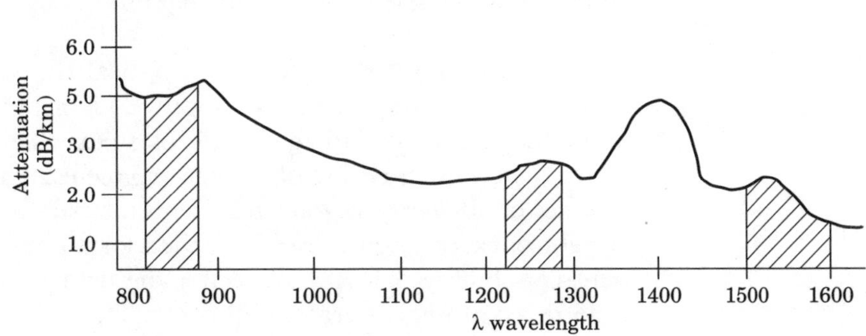

Figure 3.10, you will note three shaded areas. These areas represent common transmission windows and define the wavelengths used for most optical transmission. Thus, instead of operating at any specific wavelength, both lasers and LEDs are designed to operate at specific frequencies, resulting in specific wavelengths that minimize attenuation.

DISPERSION Although optical attenuation is an important characteristic governing the use of a fiber, it is not the only characteristic that governs performance. Another important characteristic is dispersion.

Dispersion represents the smearing or broadening of an optical signal. This phenomenon results from the components of light that consist of numerous discrete wavelengths traversing the fiber at different rates. In an optical transmission system, dispersion results in the spreading of a pulse as distance increases. Thus, dispersion limits the maximum data transmission rate or information carrying capacity of a fiber. Fiber dispersion varies with wavelengths and is denoted in terms of picoseconds (ps) per nanometer per kilometer.

Figure 3.11 depicts the relationship between dispersion and wavelength. Fiber dispersion is controlled by the fiber design process, where the location at which dispersion equals zero is referred to as the *zero-dispersion wavelength.* The zero-dispersion wavelength represents the wavelength at which a fiber's information carrying capacity peaks. For single-mode fiber the zero-dispersion wavelength is located at approximately 1310 nm. Another zero-dispersion wavelength is located in the 1550-nm region depicted by the second curve in Figure 3.11. This curve is associated with silica-based fibers and explains why lasers operating at or near 1310 and 1550 nm represent two popular light source operating wavelengths.

THE EFFECT OF DISPERSION In an optical transmission system, light pulses represent the transmission source. Light pulses are used in essentially the same way as in an electrical system, where voltage or current is applied for a given period of time to represent a binary one (1) while the absence of voltage or current represents a binary zero (0). In other words, a light pulse of given power is applied for a given time as input, representing a binary one. At the receiver the light pulses must be recognizable, or transmission errors will occur. The left portion of Figure 3.12 illustrates the transmission of light pulses into an optical fiber. Dispersion in an optical fiber results in the broadening of pulses in the time domain. Depending on the amount of dispersion, it becomes possible

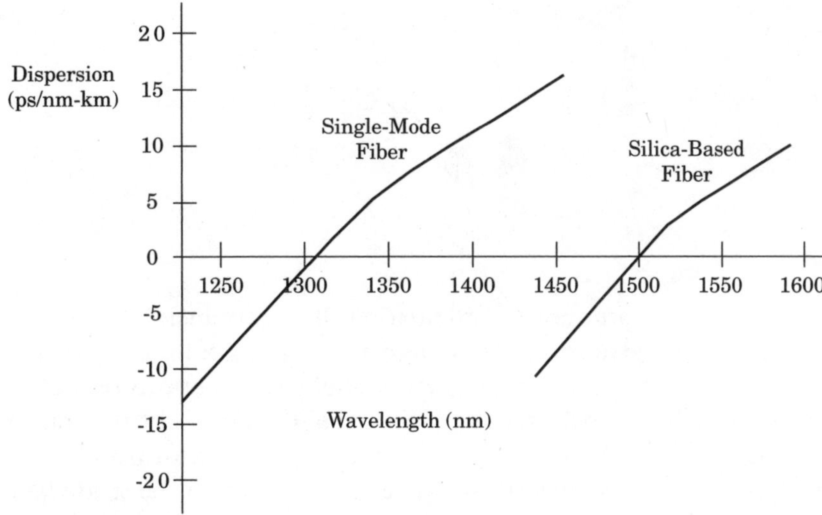

Figure 3.11 Dispersion versus wavelength.

for the trailing edge of one pulse to in effect merge with the leading edge of the following pulse as illustrated in the right portion of Figure 3.12. When too much dispersion occurs in an optical fiber, the pulse spreading can reach a point where transmission errors occur. Thus, one must either use a lower transmission rate or obtain an optical fiber that exhibits a lower level of dispersion.

Two types of dispersion affect light pulses: modal and material. As noted previously, rays of light that flow down different paths in an optical fiber reach the receiver at different times, resulting in the spreading of a pulse. This type of dispersion, referred to as *modal dispersion,* results from the differences in the propagation time of rays through a fiber.

A second type of dispersion, referred to as *material dispersion,* results from the nonlinear relationship between frequency and the index of refraction in the core of a fiber. Because different frequency components have different velocities of propagation in an optical fiber, this results in the broadening of the optical pulse.

TRANSMISSION WINDOWS In the transmission properties of glass, there is an important relationship between attenuation and the wavelength of the light source. This relationship is well known and is plotted in Figure 3.10*c*.

If you carefully examine the relationship between attenuation and wavelength shown in Figure 3.10*c*, you will note three so-called windows where transmission at certain wavelength ranges result in the lowest

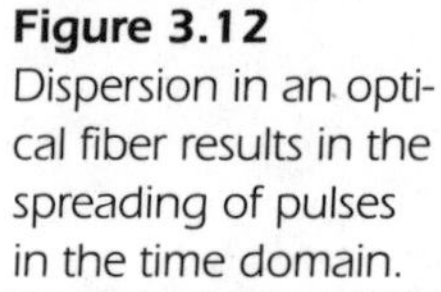

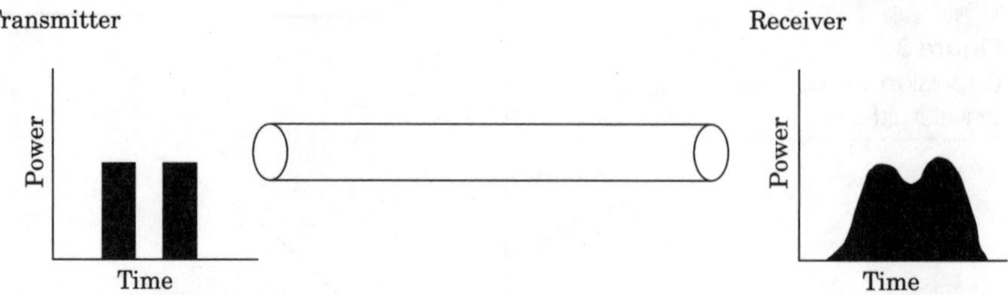

Figure 3.12
Dispersion in an optical fiber results in the spreading of pulses in the time domain.

amount of attenuation. It is extremely important to design an optical transmission system to obtain the lowest possible attenuation. This is because a low level of attenuation permits optical amplifiers to be spaced father apart, in effect reducing the cost of the transmission system.

The *first generation* of optical transmission systems operated in the 800- to 900-nm range. As indicated in the shaded area in the left portion of Figure 3.10*c*, this is equivalent to the first optical window and represents the use of light-emitting diodes (LEDs). Because the lowest level of attenuation occurs at approximately 850 nm in this window, as you might expect, early optical systems used LEDs operating primarily at that wavelength.

The *second generation* of optical fiber systems use lasers operating in the 1200- to 1300-nm range. As indicated in the shaded area in the middle of Figure 3.10*c*, this represents the second optical window, in which attenuation is significantly lower than in the first window.

Although most currently installed optical transmission systems operate within the second optical window, this is rapidly changing. A third window, which is shown in the shaded area in the lower right of Figure 3.10*c* to extend from 1500 to 1600 nm, has a very low level of attenuation and is currently the preferred operating area for optical fiber installed since 1990 or so. There are two "bands" within the range of wavelengths noted above. The C-band range is 1530 to 1570 nm, while the L-band range is 1570 to 1610 nm.

OPTICAL WINDOW UTILIZATION The use of the first optical window is for short-wavelength multimode systems. In comparison, the second optical window is used for long wavelength multimode or single-mode transmission systems.

Single-mode fiber is used primarily for long-distance carrier systems. The ITU G.652 specification for optical fiber is optimized for a 1310 nm wavelength and is referred to as *non-dispersion-shifted fiber* (NDSF). A second ITU specification, G.653, refers to dispersion-shifted fiber (DSF) and was designed for minimal dispersion in the C band. As wavelength division multiplexing (WDM) systems emerged, a defined level of chromatic dispersion was recognized to have a positive effect on the ability to place multiple light signals on a common fiber. This resulted in the promulgation of the ITU G.655 standard for *non-zero-dispersion-shifted fiber* (NZDSF). NZDSF fibers designed for long-distance carrier systems have a moderate amount of dispersion in the C and the L bands. Today the C and L bands are being used for dense wavelength division multiplexing (DWDM). In Chapter 4 we will discuss optical multiplexing as well as the use of optical amplifiers that make this technique of placing multiple signals on a fiber economically possible.

Fiber Composition

Three types of material can be used to manufacture fiber-optic cable: glass, plastic, and plastic-clad silica (PCS). The use of each type of material results in differences in the cost of the cable as well as the amount of attenuation that light pulses experience as they flow down the cable. Concerning the latter, attenuation results primarily from two physical phenomena: absorption and scattering. As light in the form of photons flows along the core of a fiber, the photons interact with molecules that make up the fiber core. This interaction results in a loss of signal energy by the photons, whose end effect is the absorption of light.

A second effect resulting from the flow of light in a fiber is scattering. As noted earlier in this chapter, the effect of scattering is to redirect a portion of light out of the core into the cladding, causing other rays to bounce from core to cladding to core as they propagate along the fiber. Because the end result is a stretched pulse, the effect of scattering or modal dispersion is to attenuate the pulse.

If we compare and contrast the three types of material used for optical fibers, we can categorize them with respect to their attenuation, cost, and utilization. Table 3.1 summarizes these properties. We will now describe and discuss these relationships by covering each type of fiber.

TABLE 3.1

Comparison of Fiber Compositions

Fiber material	Attenuation	Cost	Utilization
Glass core	Lowest	Highest	High bandwidth, long distance
Plastic core	Highest	Lowest	Low bandwidth, short distance
Glass core with plastic cladding (plastic-clad silica)	Medium	Medium	Medium bandwidth, short to medium distance

Glass Fiber

A glass-based fiber-optic cable has a glass core and glass cladding, with impurities added to obtain the desired indices of refraction necessary to guide light rays through the fiber. This type of cable has the lowest value of attenuation; however, it is also the most expensive of the three types of materials used to create an optical fiber. Concerning utilization, a glass-based fiber-optic cable is by far the most commonly used type of optical cable. Thus, it also represents the type of cable that installers are most familiar working with, which means that functions such as cable splicing and pulling techniques are well known.

Because of its low attenuation, a glass fiber is always used for single-mode optical cable. Because single-mode fiber requires a very thin case, its composition eliminates the potential use of plastic, which requires a thicker core. Another characteristic of glass fiber is its capability to support a relatively high signaling rate. Some types of single-mode step-index optical fiber can support a gigahertz signaling rate.

Plastic Fiber

Plastic-based fiber-optic cable is more durable than glass-based optical fiber. In addition, this type of optical cable is relatively inexpensive. However, it is difficult to fabricate with a thin core; as a result, the plastic core and the plastic cladding are relatively thick. Typical plastic-based fiber-optic cable is manufactured as 480/550-, 735/700-, and 9800/1000-nm cable.

Because a large core is prone to ray dispersion, plastic fiber-optic cable has the highest attenuation of the three types of material used for the construction of fiber-optic cable. Currently, this type of optical fiber is used mainly for transmission over relatively short distances, such as within a factory.

Another problem with the use of plastic-based fiber-optic cable is that it is flammable, meaning that when used within a building, applicable codes may hamper its use and require its installation within a conduit.

Plastic-Clad Silica

As the term implies, plastic-clad silica (PCS) represents an optical cable manufactured using a glass core and a plastic cladding. The glass core consists of vitreous silica, while the plastic cladding is fabricated with a lower refractive index than in the core.

The composition of a PCS-based optical cable is such that the attenuation and cost of PCS cable reside between those of a glass- and a plastic-based fiber-optic cable. A standardized PCS-based optical cable is 200/380 nm in size, with a 600-nm jacket. Although less expensive than glass, a PCS fiber-optic cable has considerable plasticity, resulting in difficulty when attempting to connect or splice cable.

Because glass core fiber provides the lowest level of attenuation and high bandwidth, it is the preferred medium for both LANs and WANs. However, at data transmission rates as high as 100 Mbits/s, both plastic core and PCS are used with LEDs in a LAN environment.

We can note the general operational characteristics of multimode and single-mode fiber by examining their general level of attenuation, operating rate, and bandwidth, as well as the general environment in which they are used. Thus, let's do so.

Today multimode fiber is used primarily in a LAN environment. This is because such fiber has an attenuation level between −3 and 1 dB/km, which is relatively high in comparison to that of single-mode fiber. The bandwidth capability of multimode fiber is between 160 and 500 MHz/km, which permits a data transmission rate of approximately 25 to 50 percent that of Gigabit Ethernet to be supported at a distance of one kilometer. This explains why, when multimode fiber is used in that LAN environment, the transmission distance is significantly less than 1 km. Although both lasers and LEDs can be used with multimode fiber in a LAN environment, less costly LEDs are used. When LEDs are used, a light source of 850 nm is employed, while the use of lasers results in a light source with a wavelength of 1300 nm.

In comparison to multimode fiber, single mode has an attenuation rate a fraction of multimode. The attenuation rate of single-mode fiber is between −0.35 and 0.20 dB/100 km, which makes it more suitable for use in a carrier's backbone infrastructure. Lasers operating at 1310 or 1550 nm are commonly used with single-mode fiber, enabling bandwidth of up to 10 GHz/100 km to be obtained today, which is expected to increase to 40 GHz/100 km in a few years. Table 3.2 compares the primary operational characteristics and utilization of multimode and single-mode fiber.

Types of Cable

Today there are literally hundreds to thousands of different types of fiber-optic cables. Some cables contain hundreds of fibers, while other cables may contain a single fiber or two, four, or more pairs of fibers.

The popularity of cable with two or more fibers dates to the early use of fiber optics. At that time a technique referred to as *wavelength division multiplexing* (WDM), which allowed multiple communications over a common fiber, was unknown. Thus, the ability to obtain two-way transmission over an optical fiber required the use of one fiber for transmission and a second for reception.

Figure 3.13 illustrates how a pair of fibers are used to obtain bidirectional transmission capability. Note that at each end of the cable a transmitter in the form of a light source and a receiver in the form of a light detector are connected to each fiber in the cable.

The cable shown in Figure 3.13 consists of two fiber-optic cables, each containing a single fiber, encased in a common jacket. Over the years numerous types of fiber-optic cables have been manufactured that build

TABLE 3.2

Comparison of Operational Characteristics and Utilization of Multimode and Single-Mode Fiber

Characteristic	Multimode Fiber	Single-mode Fiber
Application	LAN	WAN
Attenuation	−3 to −1 dB/km	−0.35 to −0.20 dB/km
Wavelength	850/1300 nm	1310/1550 nm
Light Source	LED, laser	Laser
Bandwidth	160–550 MHz/km	10 GHz/100 km

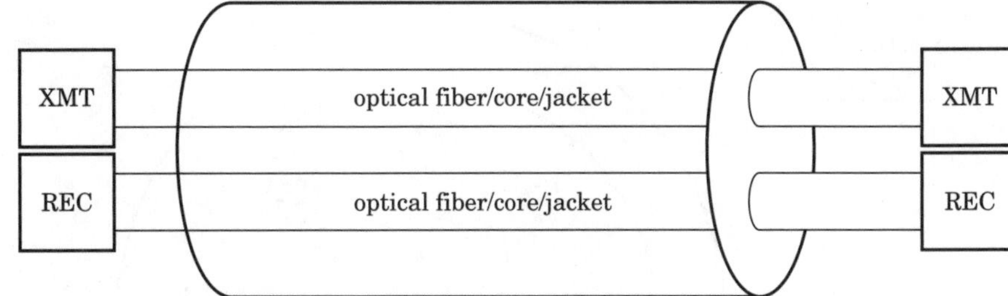

Figure 3.13
Most fiber-optic cables include one or more pairs of fibers, with each pair supporting two-way full-duplex communications.

on this design. Some cables contain two, four, six, eight, or more individual fiber-optic cables, each containing one or more fibers.

Cable Architecture

One example of the cable architecture described above is shown in Figure 3.14. In this example a common cable housing holds eight cables, with each cable containing eight fibers. Thus, this cable provides 8 × 8, or 64 optical fibers. By increasing the number of fibers per cable and the number of fiber-optic cables within a common cable jacket, it becomes possible to rapidly increase the total number of fibers. Because the cost of cable installation can easily exceed the cost of the cable, most communications carriers install a large number of fibers even if their initial requirement calls for the use of just a few. As requirements increase, the carrier will use additional fibers in the bundle. The term "dark fiber" is associated with unused fiber, while the term "lighted fiber" refers to fiber in use.

In addition to the number of fibers within a cable, the bundled cable can vary according to the material used in its construction. Some individual and bundled cables are designed for general-purpose outdoor use, while other cable is designed for indoor use. In addition, you can obtain *tactical cable* designed for ruggedness required by the military forces as well as other types of specialized cable. Thus, there are literally hundreds of cables available for different applications.

Now that we have a general understanding of the types of cable used, we will conclude this chapter by discussing another important cable-related topic: the fiber-optic connector.

Figure 3.14
Multiple-fiber cable.

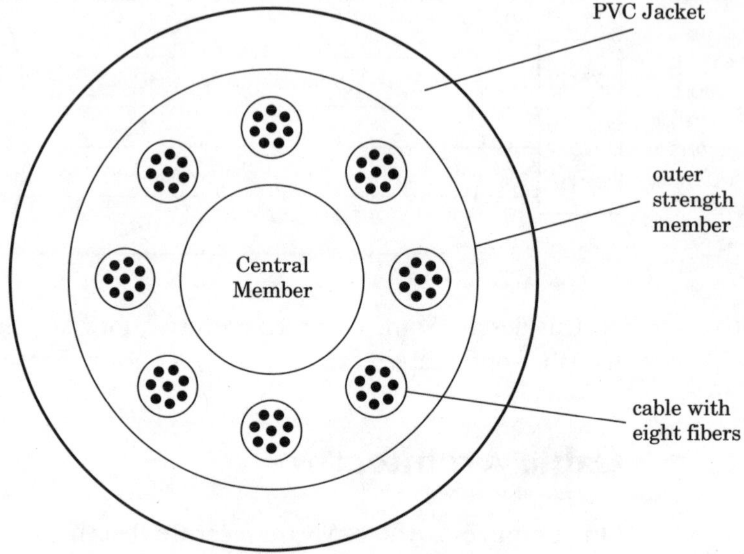

This fiber bundle provides eight fiber cables, each with eight fibers, resulting in 64 fibers becoming available for use.

Connectors

A *connector* represents an important component of a fiber-optic transmission system. In addition to terminating cables, connectors are used to route an optical signal and represent an important component used for the cable reconfigurations required to satisfy changing organizational or customer requirements.

The connector is a mechanical device physically connected to the end of a fiber-optic cable. This device is designed to mate with another device to provide attachment between cables, from a cable to a light source, or from a cable to a receiver. The attachment mechanism is referred to as the *coupling method* and warrants discussion. Because the coupling method is related to one or more connectors, we will examine both as a single entity.

Coupling Method and Connectors

Three basic types of coupling methods are used by connectors: threaded, bayonet, push-pull.

THREADED CONNECTOR The threaded type of connector is used by a submulti assembly (SMA) connector. The SMA connector was introduced by Amphenal Corporation and included a threaded cap and housing. This is a first-generation connector whose use has considerably decreased in favor of ST (straight tip)- and SC-(stick and click)-type connectors.

One problem associated with the use of a threaded connector is the airgap caused by a poor connection. Because a poor connection results in the medium changing from fiber to air and back to fiber, this results in a change in the index of refraction. This change results in a *Fresnel reflection loss,* in which light is reflected back toward the transmitter.

BAYONET CONNECTOR A second type of connector coupling results in the use of a keyed bayonet similar to a bayonet coaxial (BNC) connector, a male connector mounted at each end of a coaxial cable (used in Ethernet configurations). This type of connector requires a twist-turn fastening operation, which reduces the possibility of an airgap formation commonly associated with a threaded connector, which lacks a control for rotational alignment.

Two popular keyed bayonet-type connectors are the ST (straight tip) and ST-II connectors pioneered by AT&T during the mid-1980s. Both are used with single-mode and multimode fiber and are keyed and spring-loaded, respectively. Connections are established via a push-in and twist movement.

PUSH-PULL CONNECTOR A third coupling method can be categorized as push-pull. One popular push-pull connector is the SC connector, which includes a locking tab. This connector is used primarily with single-mode fiber optic cable and uses a ceramic ferrule. The *ferrule* is a long, thin sleeve in which a fiber is placed. The ferrule acts as an alignment mechanism and represents a critical component of the push-pull coupling method.

PLASTIC FIBER CONNECTORS It is common to encounter connectors such as SMA and ST, which were developed for glass fibers used with plastic fiber-optic cable. In addition to the use of SMA and ST connectors, some manufacturers use proprietary designs. The latter obviously restricts interconnection capability. Table 3.3 provides an alphabetical list of common connectors used with glass fiber-optic cable. This list should not be viewed as all-inclusive as many additional proprietary connectors have been introduced by vendors since 1990.

Insertion Loss

The insertion of a connector results in a loss of optical power. We can determine that loss, which is referred to as *insertion loss,* by measuring power before and after the insertion of a connection. Thus, the insertion loss becomes

$$dB = 10 \log_2 \frac{P_1}{P_2}$$

TABLE 3.3

Connectors Commonly Used with Glass-Based Fiber-Optic Cable

Connector type	Description
Biconic	Uses a tapered sleeve fixed to the fiber-optic cable; when the plug is inserted into a biconic receptacle, a tapered end provides a mechanism for locating the fiber-optic cable in an appropriate position (although one of the earliest connector types used with fiber optics, this cable is little used today)
D4	Originally designed by Nippon Electric Corporation, the D4 is similar to the FC connector; it uses threaded coupling and a 2.0-mm-diameter ferrule
FC	The FC connector uses a position locatable notch and a threaded receptacle to provide extremely precise positioning for single-mode fiber-optic cable
FCPC	Uses a spring-loaded full-ceramic ferrule
LC	A Lucent developed small form connector that fits into a standard RJ-11 jack; LC supports both single- and multimode fiber
SC	Employs a push-pull coupling method with a locking tab; the SC connector is used mostly with single-mode fiber-optic cable
SMA	Uses a threaded cap and housing; considered the predecessor of the ST connector (the SMA connector has been rapidly replaced by the use of ST and SC connectors)
ST	Uses a keyed bayonet coupler similar to a BNC connector (the ST connector is in widespread use and can support both single-mode and multimode fiber); there are two versions, referred to as ST and ST-II—these are keyed and spring-loaded, push-in and twist-type connectors

where P_1 is the power measured before insertion of a connector and P_2 is that measured after insertion. SMA connectors, which represent the first generation of connection products, have insertion losses ranging from 0.60 to 0.80 dB. The second generation of connectors, represented by biconic and ST types, have an insertion loss of approximately 0.50 dB. Third-generation connectors, such as the LC and SC, have an insertion loss as low as 0.20 dB.

The EIA/TIA 568 Standard

The performance requirements for installed optical fiber within a building were addressed by the Electronic Industry Association (EIA) and the Telecommunications Industry Association (TIA) in a joint publication, the EIA/TIA-568-B.1 *Commercial Building Telecommunications Cabling Standard.* This standard defines both the maximum level of optical attenuation and the maximum transmission distance for horizontal, backbone, and centralized optical fiber cabling connections. As noted earlier, optical attenuation represents a reduction in light power or intensity as it flows through a cable, cable connections, and splices. Thus, optical attenuation represents insertion loss on an end-to-end basis.

The actual attenuation permitted under the EIA/TIA 568-B.1 specification depends on several factors, such as the type of fiber used, the number of connections used to mate one fiber to another, the number of splices, the optical wavelength of the transmitter, the cable length and whether cabling is horizontal, backbone, or centralized. *Horizontal cabling* represents the cabling between the telecommunications outlet in a work area and the horizontal cross-connect. In comparison, *backbone cabling* represents cabling between telecommunications rooms, while centralized cabling represents cabling from a work area to a centralized cross-connect via the use of pull-through cables, a splice, or an interconnect in a telecommunications room.

Table 3.4 summarizes EIA/TIA 568-B.1 link loss and length for multimode 62.5/125-μm and multimode 50/125-μm fiber. As indicated by the reference to two equations in the footnotes of the table, the maximum loss for centralized and backbone cabling is based on the number of connection pairs, the number of splices, and the length of the cable.

TABLE 3.4

EIA/TIA-568-B.1 Multimode 62.5 and 50/125-μm Optical Fiber Link Loss and Cable Length

	Maximum loss		
Cable Use	**850 nm**	**1300 nm**	**Maximum length, m**
Horizontal	2.0 dB	2.0 dB	90
Centralized	See Eq. 1*	Eq. 2†	300
Backbone	See Eq. 1*	Eq. 2†	2000

*Equation 1:
Maximum loss = (number of connection pairs × 0.75 dB) + (number of splices × 0.3 dB) + (length × 3.5 dB/km)

†Equation 2:
Maximum loss = (number of connection pairs × 0.075 dB) + (number of splices × 0.3 dB) + (length × 1.5 dB/km)

In examining the footnotes in Table 3.4, you will notice values associated with a mated connection (connector pair), splice, and cable length. Those are the recommended values for each category of impairment that generates an optical loss. Thus, you can compute a relevant loss budget by determining the length of the link, the number of connectors, and the number of splices and then use the relevant equation.

CHAPTER 4

Light Sources and Detectors

By itself, an optical fiber can be considered as equivalent to a hose without water or a wire without current. To provide the system with transmission capability, an optical fiber requires a light source at one of its ends and a light detector at the destination. Thus, the purpose of this chapter is to become acquainted with the use of optical fiber as a transmission system by focusing on the transmitter and the receiver used in such systems. To accomplish this goal, we will discuss the operating characteristics of light-emitting diodes (LEDs) and lasers. Once this is accomplished, we will consider the operational characteristics of different types of optical receivers, which are commonly referred to as *detectors* or *optical detectors*. Because you cannot simply shine light into a fiber or place a detector at the end of a fiber, the roles of a coupler and its connector are extremely important. Thus, we will also cover the use of these components in this chapter. However, before beginning our coverage of light sources and detectors, let's make sure that we understand how they are employed. Thus, let's begin this chapter by focusing on the components of an optical transmission system.

Components of an Optical Transmission System

An optical transmission system in many ways is very similar to other types of transmission systems, such as an electrical or a wireless transmission system. Each type of transmission requires three key components: a transmitter, a transmission medium, and a receiver. Figure 4.1 illustrates the relationship among the three components.

Transmitter and Receiver

In an optical transmission system both the transmitter and the receiver are electricooptical devices. Thus, they either convert an electrical signal into an optical signal (transmitter) or convert an optical signal into an electrical signal (receiver).

Medium

In an optical transmission system, the medium represents an optical fiber. That fiber can be a single-mode or multimode glass fiber or a plastic multimode fiber. Because we already examined the transmission

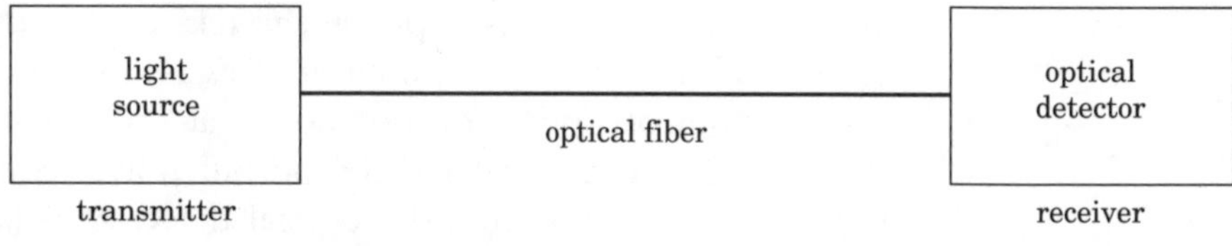

Figure 4.1 The key components of an optical transmission system include a transmitter in the form of a light source, the optical fiber that represents the transmission medium, and a receiver in the form of an optical detector.

medium in Chapter 3, readers are referred back to that chapter for information about optical fiber as a transmission medium.

Coupler

The connection of a fiber to a light source and light detector is not a simple process. Instead, a special coupler must be used to focus light into a fiber and at the opposite end focus light flowing out of the fiber to a light detector. Although not shown in Figure 4.1, the coupler, as well as the type of connector on the coupler, play important roles in creating an optical transmission system and are covered later in this chapter.

Component Relationships

An optical transmission system is similar to an electrical transmission system in that the power at the receiver must be equal to or above its sensitivity level to be recognized. By examining the power level of the transmitter, the receiver sensitivity level, and the attenuation level of couplers and connectors, we can determine the transmission distance possible prior to amplification becoming necessary. This may not appear important when installing fiber within the floor of a building; however, it becomes an important consideration for communications carriers as it provides them with the ability to determine the number of amplifiers required on a long-haul circuit. Because each amplifier requires power and shelter, the ability to space amplifiers farther apart from one another will affect the initial and recurring cost of a long-haul fiber circuit. To illustrate a simplistic example of the relationship of components affecting the distance between amplifiers, let's assume that we are using a laser with 0 dBm of optical output power at the 1550-nm wavelength. As a refresher, 0 dBm means that the output power equals the input power. Let's further assume that the optical detector to be used has an input sensitivity of −20 dBm. Again as a refresher, −20 dBm means that the receiver is capable of detecting one-hundredth of the optical power injected into the fiber by the transmitter.

Because we transmit at 0 dBm and have a receiver sensitivity of −20 dBm, the long-haul fiber-optic system becomes capable of supporting an attenuation of 20 dBm of light. Let's further assume that the optical fiber has an attenuation of 0.20 dBm/km at 1550 nm and the connectors at both ends of the cable have an attenuation of 0.10 dBm. From the preceding observations, we can compute the long-haul drive distance before optical amplification becomes necessary as follows.

First, we would reduce the 20 dBm of allowable light by 0.10 dBm × 2 for the couplers on each end. Thus, we have 19.8 dBm of adjusted allowable attenuation. Since the cable attenuation is 0.20 dBm/km, this means that the permissible long-haul distance becomes 19.8 dBm, which at 0.20 dBm/km, results in 99 km.

Now that we have an appreciation for the major components associated with an optical transmission system and how they can govern the obtainable transmission distance before optical amplification becomes necessary, let's devote the remainder of this chapter to identifying the primary components of an optical transmission system other than the optical cable that was discussed in Chapter 3.

Light Sources

As mentioned earlier in this chapter, two basic types of devices are used to convert electronic signals into light signals for fiber-optic transmission: light-emitting diodes (LEDs) and lasers. In this section we will examine each device separately.

Light-Emitting Diodes

The *light-emitting diode* (LED) represents a semiconductor device that emits incoherent optical radiation in the form of electromagnetic waves when biased in the forward direction. Optical radiation in the form of light is referred to as "incoherent" as it radiates in all directions from the surface of an LED. This is illustrated in Figure 4.2, which shows how light generated by an LED flows in many directions.

Although we will discuss LED operation in detail in the next subsection, we can note several interesting general operational characteristics of an LED-based optical transmission system: (1) because of the dispersion of light, we need to use a thicker core fiber; (2) because of light dispersion, it is extremely important to attempt to couple the LED as best as possible to the fiber to minimize light loss; and (3) because of the incoherent optical radiation, the transmission distance associated with LEDs is considerably less than that obtainable with lasers. Nevertheless, LEDs are relatively inexpensive and serve well for short transmission distances. This explains why most early fiber-optic transmission systems were based on the use of LEDs as a light source. Operating at a wavelength of 850 nm for use with multimode optical fiber, early LED-based systems were capable of transmitting at rates of up to 140 Mbits/s. Unfortunately, such systems required repeaters every 10 to 15 km. With the development of low-cost lasers and the migration to single-mode fiber by most communications carriers, the use of LEDs is now relegated primarily for LAN and other short-distance applications. Today the upper throughput limit of LED-generated signals is approximately 300 Mbits/s. In comparison, lasers routinely operate at 10 Gbits/s and in field trials now operate at 40 Gbits/s. Now that we have a general appreciation for LEDs versus lasers, let's examine how the former operates.

Figure 4.2 Light generated by an LED is incoherent and spewed in all directions.

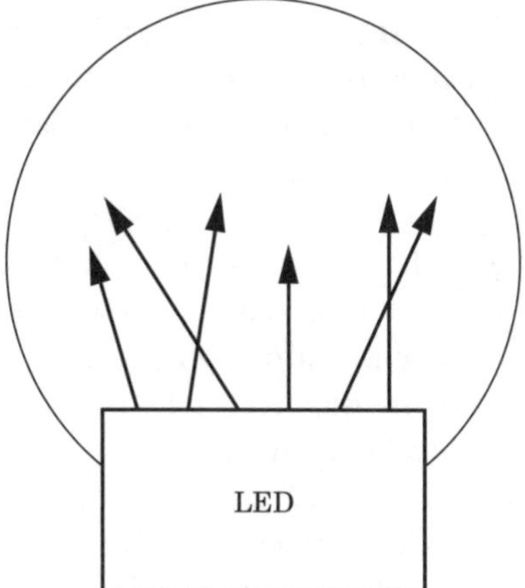

Operation

Light can be generated in several ways. The familiar lightbulb represents incandescent light since optical radiation results from the temperature applied to a filament. In comparison, an LED represents light generated by the electronic excitation of a material, a term referred to as *luminescence.* To understand how semiconductor chips, including LEDs and lasers, operate, let's review some basic physics, including the structure of atoms and the p-n semiconductor junction, to see how a semiconductor becomes capable of generating light.

THE COMPOSITION OF AN ATOM Figure 4.3 illustrates the composition of an atom. An *atom* consists of an inner nucleus surrounded by electrons that circle the nucleus. That circle is referred to as a *shell.* Each shell has a maximum number of electrons, ranging from two in the inner shell (K) to 32 in the outer shell (N). The outer shell is referred to as the *valence shell* or *valence band,* and represents the shell that is involved in chemical bonding, which permits compounds to be developed.

The ability of an electron to flow as current depends on the molecular structure of a material, including its valence shell. Specifically, an electron must break its bond to an atom to obtain the ability to flow as current. When it breaks its bond, the electron is said to be in the *conductive band.*

CONDUCTORS The degree of movement of an atom from a valence band to the conductive band determines whether the material is a good or poor conductor. For example, copper, gold, and silver have only one valence electron. This means that those elements require a minimal amount of energy for an electron to break its band. Thus, this explains why those materials are relatively good conductors.

THE p-n JUNCTION In the world of electronics the semiconductor *positive-negative* (p-n) *junction* represents the basic material used to develop LEDs and lasers. Because p-n junction operation is easily described in terms of the use of silicon material, let's continue our discussion of atoms by focusing on the silicon atom.

A *silicon atom* has four valence electrons. Those electrons form the bonds that hold silicon atoms together, which in turn forms the crystalline structure of silica. The silicon bonds are also referred to as *cova-*

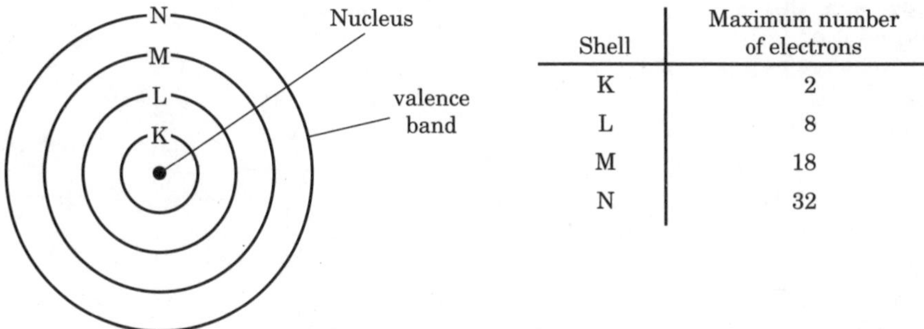

Shell	Maximum number of electrons
K	2
L	8
M	18
N	32

Figure 4.3 The electronic structure of an atom.

lent bonds because the atoms share their electrons. Figure 4.4 illustrates the covalent bonds in a series of silicon atoms. Note that each silicon atom has access to eight valence electrons: four of its own and four from surrounding atoms. All the electrons are covalent bonds, and initially none are available as free electrons.

THE FREE ELECTRON AND n-TYPE MATERIAL To see how a p-n semiconductor junction operates, assume that an atom with five valence electrons is added to a silicon crystal. Since only four electrons are required for the covalent bond, this results in a free electron. Assume that this electron represents a free electron in the covalent band that can move around the structure. Because the material now has an excess of negatively charged electrons, it is also referred to as an *n-type material* and represents one-half of a p-n semiconductor. As you might expect, the other half of the p-n semiconductor will have an excess of positively charged material.

p-TYPE MATERIAL To illustrate the creation of material with an excess of positively charged electrons, let's assume that other material with a composition of only three valence electrons is added to silicon. Because four valence electrons are required for a covalent bond, one would now be missing. This electron vacancy is referred to as a "hole" in the form of a positively charged particle. Since the resulting material has an excess of positively charged holes, it is referred to as *p-type material.*

DOPING MATERIAL The p- and n-type material described above is created by doping the material with atoms. Since the 1970s, gallium arsenide (GaAs), gallium arsenide phosphide ($GaAs_{1-x}P_x$), aluminum gallium arsenide ($Al_xGa_{1-x}As$), aluminum gallium indium phosphide

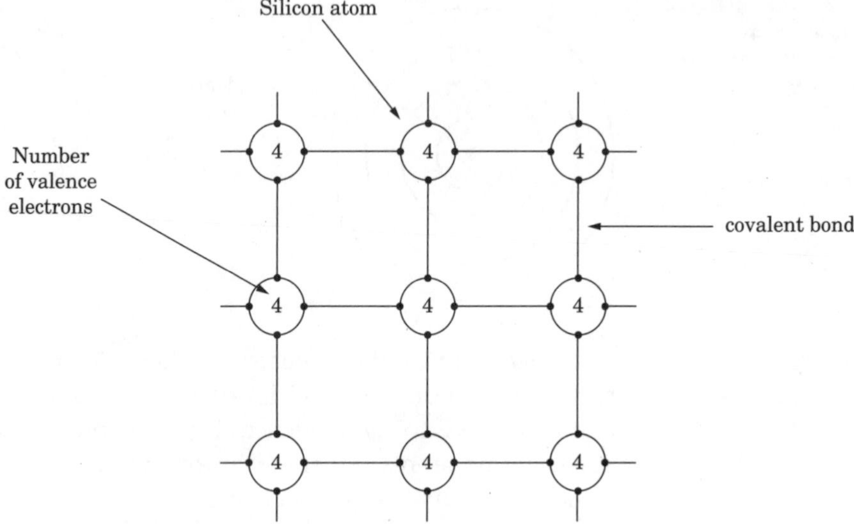

Figure 4.4 Covalent bonds in a series of silicon atoms.

$[(Al_xGa_{1-x})_yIn_{1-y}P]$, and gallium indium arsenide phosphide ($Ga_xIn_{1-x}As_yP_{1-y}$) (where *x* represents a percent of the elements that are aluminum and 1-*x*, the percent that are gallium) have been used as doping material. Such material represents elements from groups III and V of the periodic table that have three and five electrons in their valence shells. Through the use of elements from groups III and V of the periodic table, it becomes possible to create a p-n semiconductor that generates energy in the form of light. In comparison, silicon generates energy in the form of heat resulting from the vibrations in the crystal structure of silica.

ELECTRON FLOW By joining p and n material, holes and electrons flow across the p-n junction and recombine. Here the process of recombining means that a free electron falls into a hole as it moves from the conduction band to the valence band, with both the hole and the electron losing their charge. Because of the recombination of holes and electrons around the p-n junction, a barrier forms as the charges are depleted, which prevents the additional migration of electrons and holes across the junction. The only method suitable to enable additional migration across the p-n junction is to apply energy to the material. As energy is applied, this action results in holes and electrons continuing their recombination, and energy is emitted. When applicable material is used, the energy emitted will be in the form of light, providing for creation of the LED. Thus, an LED represents a p-n semiconductor with applicable doping to emit light when a forward bias is applied to the semiconductor.

Figure 4.5 generally illustrates the application of a positive voltage across an appropriately doped p-n semiconductor to generate light. Note that the term *forward bias* in Figure 4.5 refers to the connection of the negative-battery terminal to n-type material and the positive-battery terminal to the p-type material. This action results in the injection of electrons into the n-type material and the creation of electrons in the p-type material by the extraction of electrons from that material. As a result of this electrical action, both electrons and holes move toward one another across the depleted area of the junction. As they combine, their energy is emitted in the form of light as shown in the lower portion of Figure 4.5. Because current must be continually supplied to maintain an excess of carriers for recombination, the toggling of the forward bias on and off controls the generated light. Thus, by varying the forward bias on and off, you are now able to generate light and turn it off. By varying the forward bias in tandem with the value of electrical pulses, it becomes possible to convert electrical signaling into optical signaling.

One of the key limitations of LEDs is the optical pulse rise and fall times. Because light is not instantly turned on when the forward bias becomes operative, there is a delay in obtaining an applicable light pulse. Similarly, there is a second delay between toggling the forward bias off and the light pulse attenuating to black. As a result of these delays, the maximum achievable LED data transmission rate is approximately 300 Mbits/s.

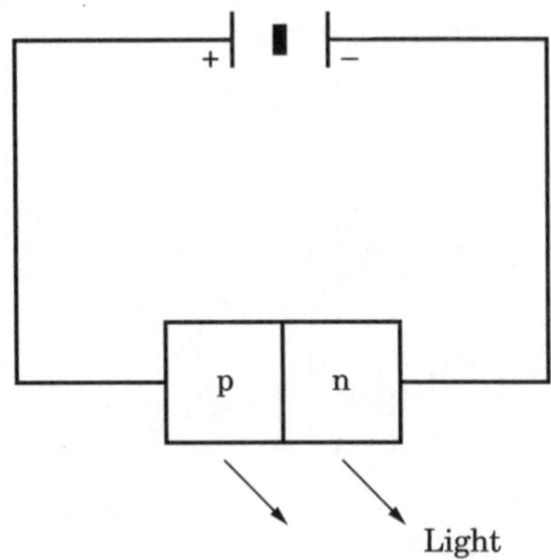

Figure 4.5
A light-emitting diode is a p-n semiconductor that emits light when a forward bias is applied.

OPERATIONAL RANGE The ratio of the number of emitted photons to the number of electrons that cross the p-n junction represents the *quantum efficiency* of an LED. While the quantum efficiency of an LED is important, another important characteristic is the wavelength of emitted light. LED emissions occur in the visible part of the spectrum with wavelengths ranging from 400 to 700 nm or in the near-infrared band with wavelengths between 2.0 and 0.7 nm. Some commonly available LEDs operate at wavelengths of 565 nm and generate green light, and some operate at 560 nm and generate red light. In fact, LEDs that operate at 660 nm are commonly used for panel readouts that generate information at sporting events and similar functions.

Figure 4.6 is a cross-sectional view of a surface LED. Here the term *surface* indicates that light is emitted from the surface of the LED. Note that the doping material for this particular type of LED is indium gallium arsenide phosphorous (InGaAsP) and indium phosphorous (InP). Also note the use of silicon dioxide (SiO_2) as an antireflection coating along the surface of the diode.

Because an LED generates light over a wide angle, the ability to direct light into a fiber is difficult. In fact, LEDs have a coupling efficiency of approximately 2 percent, even when a coupler with a focusing lens is employed to direct light from the LED into an optical fiber. In one technique for increasing the coupling efficiency of LEDs to optical fiber, the LED is bonded to a fiber using epoxy resin. By eliminating a connector,

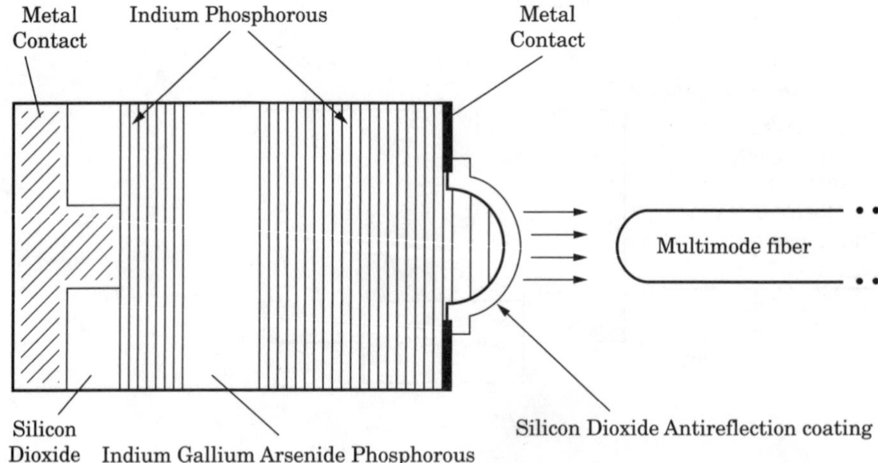

Figure 4.6 A cross-sectional view of a surface LED. Light is emitted over a broad spectrum, and even the use of a focusing lens results in a coupling efficiency of approximately 2 percent.

it becomes possible to eliminate a small portion of attenuation associated with coupling of an LED to fiber.

For short-range fiber-optic transmission at distances of up to 100 m, you can use visible light generating LEDs with either glass or plastic fiber. For longer-range transmission it is important to use an LED that generates light that has the lowest level of attenuation in dB/km for the fiber to be used. This results in infrared LEDs becoming more suitable than LEDs that generate visible light. Thus, LEDs with a wavelength between 800 and 900 nm or above 1000 nm are commonly used for transmission beyond distances of 100 m. The selection of a particular LED wavelength is commonly related to the optical windows associated with the properties of optical cables. As we noted in Chapter 3, the relationship between attenuation and wavelength of optical signals is such that there are three wavelength bands or windows where attenuation is minimal. By using an optical transmitter functioning within a window, it becomes possible to minimize optical attenuation, which, in turn, enables longer cable runs before optical amplification becomes necessary. The most popular optical window used by LEDs is currently the first one that occurs at wavelengths between 800 and 850 nm. While LEDs operating at lower wavelengths are still used for relatively short distances, at distances beyond 100 m the use of an 850-nm LED is most popular.

COUPLING LOSS One important limitation associated with all LEDs regardless of the type of light emitted is the coupling loss associated with mating the LED to an optical fiber. Because an LED emits incoherent light over a broad spectrum, its coupling loss is higher than that when a laser is used as a light source. Thus, the spectral width governs the effective signal bandwidth since a larger spectral width requires a larger portion of the optical fiber bandwidth. This is illustrated in Figure 4.7, where the bandwidths of a typical LED and laser diode are compared.

The area under each curve shown in Figure 4.7 represents the optical power generated by each device. Although our first inclination is to note that the LED generates more power than a laser diode does, as a famous radio announcer would say, "That's part of the story." This is because the spectral width, which is the half-power spread, results in a much higher coupling loss when an LED is used. In an attempt to limit coupling loss, LEDs may include a focusing lens bonded to the semiconductor and, as noted previously, can also result in the bonding of the LED to a fiber. However, for most applications the use of LEDs requires a

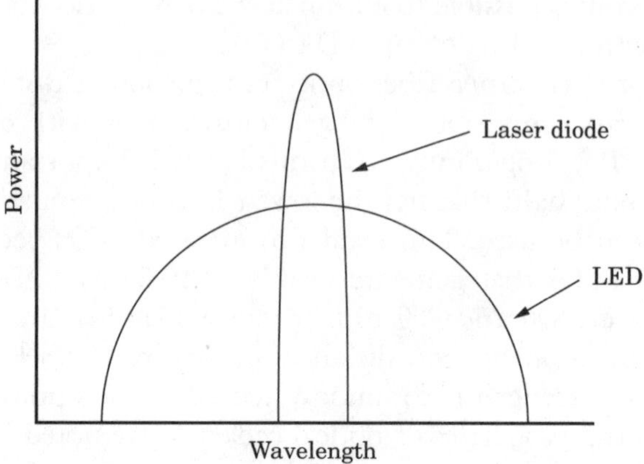

Figure 4.7
Comparison of the spectral width of an LED and laser diode.

thicker diameter fiber and results in a maximum drive distance of a few kilometers. In addition, because the LED does not instantly generate and stop producing light in tandem with applying and removing a forward bias, the obtainable data transmission rate is also limited. LEDs are commonly used for data rates under 20 Mbits/s even though they can operate at a data rate up to approximately 300 Mbits/s.

Modulation Methods

As with analog modems, more than one method can be used to convey information via an optical fiber. Different methods that were developed for the modulation of information include intensity modulation, frequency-shift keying, phase-shift keying, and polarization modulation. Because of the short drive distance associated with the use of LEDs and the simplicity of intensity modulation, this technique is used by the transmitter section of an LED that receives an electrical signal for conversion to an optical signal.

Under intensity modulation the light signal is turned on and off to represent binary values of 1 and 0. The modulated carrier is given by

$$E_s(t) = E_0 m(t) \cos(2\Pi f_s t)$$

where E_s = energy level of the output signal
E_0 = energy level of the input signal

$m(t)$ = the modulating signal
f_s = the optical carrier frequency
t = time

Because optical power is emitted in proportion to the forward bias, this power will take the shape of the input current. Thus, when the input current has a waveform of $m(t)$ that represents binary information, the resulting optical signal will resemble a burst of light when $m(t)$ represents a 1 and will be absent when $m(t)$ represents a 0. While this translates to the presence and absence of light, it is also worth noting that the linear relationship between input current and light output makes it possible to use LEDs with an analog input. When used with an analog input, an LED can typically support modulation rates of up to 20 MHz.

Comparison between an LED and a Laser Diode

In concluding our discussion of LEDs, let's compare their operational characteristics to those of a laser diode. Although we discuss the latter in the next subsection, we can note that the LED has a high coupling loss that results in less optical power flowing into a fiber. In addition, because of the longer pulse rise and fall times, the LED operates at a lower data transmission rate than does a laser diode. Because of the greater optical dispersion of the LED, it is used with multimode fiber. Although these factors limit the use of LEDs, it is also important to note that they are relatively inexpensive in comparison to a laser diode. In addition, they have a higher reliability and use multimode fiber, which can be 3 to 5 times less expensive than single-mode fiber. Hence we can paraphrase Mark Twain and say that the death of LEDs for use in optical networking is greatly exaggerated.

Lasers

A second device used as a light source with optical fiber is the laser. An acronym for *light amplification by stimulated emission of radiation,* a *laser* represents a coherent light emitter. Here the term "coherent" means that all emitted photons travel in the same wave pattern, similar to the phenomenon that occurs when a stone is dropped into a pond, generating a series of waves traveling in a circular direction away from the location where the stone hit the water.

Several types of lasers have evolved since their discovery during the 1960s; the major difference is the energized substance that increases the intensity of light that passes through the device. That energized substance, illustrated in Figure 4.8, is referred to as an *amplifying medium.* That medium can be a solid, a liquid, or a gas.

AMPLIFYING MEDIUM In the world of communications the amplifying medium is almost always a thin layer of semiconductor material sandwiched between other semiconductor layers. The resulting laser is referred to as a *laser diode,* and the amount by which the intensity of light is increased is referred to as the *gain* resulting from the amplifying medium. That gain depends on several factors, including the wavelength of incoming light, the length of the amplifying medium, and the degree to which the amplifying medium was energized. Thus, the gain can be considered to be variable. Because light amplification is an essential function of a laser, let's focus on this process.

LIGHT AMPLIFICATION As noted earlier in this chapter, the outer orbit of an atom contains electrons that can be easily raised to a higher energy state. Let's assume that a photon of light is absorbed by an atom in which a hole in the n-type semiconductor material is initially in a low energy state denoted by a 0. In comparison, we can raise the energy of an electron in the p-type semiconductor material to a higher energy state, which we can denote as 1. The electron will remain in this excited state for a very short period of time and then return to the lower state of 0 as it emits a photon of light.

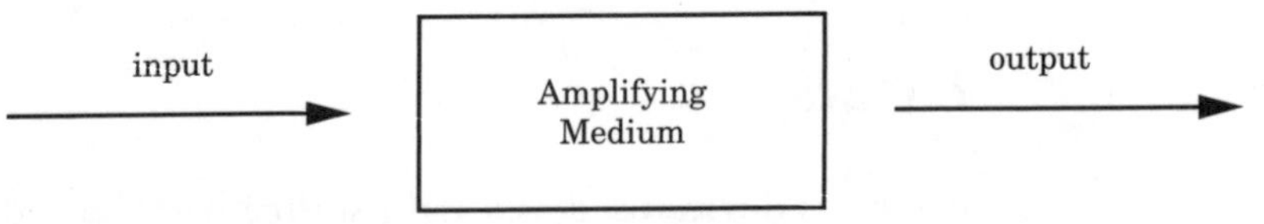

Figure 4.8 All lasers consist of an energized substance that increases or amplifies the intensity of light passing through it. That substance can be a solid, a liquid, or a gas.

SPONTANEOUS EMISSION The top portion of Figure 4.9 is a simplified view of the change in the state of an electron. The process in which an electron falls to the lower energy level is referred to as *absorption*. As the electron is absorbed, it spontaneously emits a photon of light. Thus, the movement of electrons is also referred to as *absorption with spontaneous emission.*

The actual absorption process results in the energy of the photon being equal to the difference between the two energy levels, which in our simplified example are levels 0 and 1. Because photons have different energy levels specified in electronvolts, which correspond to their wavelength, this also means that only photons of a particular frequency (or wavelength) will be absorbed. Because the best result that can be achieved under spontaneous emission is for one photon to be generated for each photon that is absorbed, this process does not result in the amplification of light.

STIMULATED EMISSION Instead of spontaneous emission, a laser must perform stimulated emission. To do so requires a photon of light to interact with an atom in an excited state, stimulating the return of the electron to the lower state and causing two photons to be emitted. A

Figure 4.9
Spontaneous versus stimulated emission.

a. Absorption and spontaneous emission.

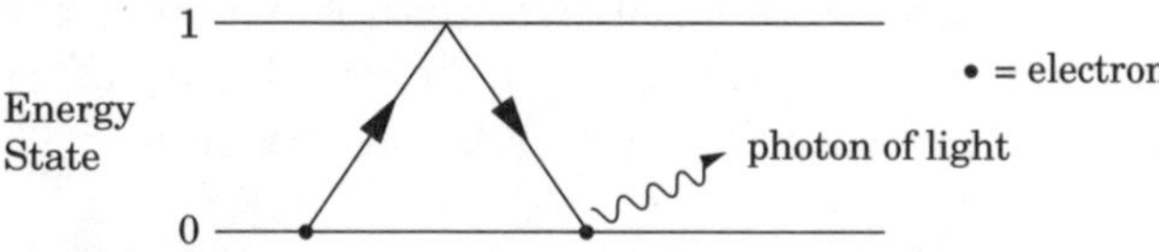

A photon absorbed by an atom is initially in a low energy state (0). As the energy of the electron is raised (1) it remains in an excited state for a short period of time and spontaneously returns to the lower state (0) where it emits a photon of light.

b. Stimulated emission.

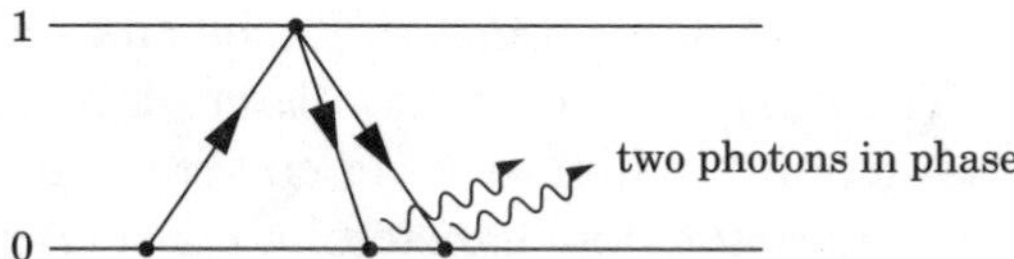

If an electron is excited and interacts with a previously excited atom, two photons, each in phase with the other, result in the generation of coherent light.

side effect of the pair of photon emissions is the fact that they are in phase with one another, which results in the generation of coherent light. This is illustrated in Figure 4.9*b*.

The coherent light generated by a laser represents a very narrow beam. That beam is normally sharply monochromatic, which means that it occupies a single color or frequency; however, it can also be generated with more than one monochromatic frequency. When the latter occurs, the use of filters enables a single frequency to be separated from multiple monochromatic emissions.

THE LASER DIODE The semiconductor laser, commonly referred to as a *laser diode*, uses a p-n junction that generates light similar to the way an LED emits light when a voltage is applied and current flows. However, unlike an LED, a laser diode requires the recombination of photons to be confined across the junction area and reflected by the use of partially reflecting surfaces that result in the formation of a cavity. Parallel mirrors are formed through the use of groups III and V compound semiconductors that can be cleaved along their planes. Those mirrors confine injected electrons and light generated by photons within a region where they interact to enhance the stimulated emission process. Thus, a key difference between an LED and a laser diode is the population inversion and optical feedback of the semiconductor laser. Through optical feedback a laser allows photons generated by recombination to stimulate additional electron-hole recombinations. By the use of higher levels of doping and an increased level of current, additional excited electrons result in a population inversion.

SINGLE- VERSUS DOUBLE-HETEROJUNCTION LASER DIODE The actual junction in a single crystal between two dissimilar semiconductors is referred to as a *heterojunction*. This type of junction is required to confine light to the region of the p-n junction as well as to obtain the ability to continuously operate at room temperatures. The use of a heterojunction (also referred to as a *heterostructure*) resulted in two general types of laser diodes: single heterojunction and double heterojunction. The single-heterojunction laser diode has a high output power, however, it requires a high driving current and has a limited duty cycle. In comparison, the double-heterojunction laser diode is not limited by its duty cycle. While the ratings of most double-heterojunction laser diodes are between 3 and 10 mW, their optical power can range up to 20 mW. The wavelengths of both types of laser diodes range from 750 mn at the highest near-infrared energy level down to 390 nm at the lower ultraviolet energy level. Many laser diodes have a wavelength near 900 nm and a rise time to achieve full output between 0.1 and 2 ns.

In the evolution of fiber-based networks, the use of lasers at 800 nm represents a first generation of optical networks. Laser systems now commonly operate at approximately 1300 and 1500 nm, which represents two low-loss windows in fiber where the attenuation level is minimal. As we noted earlier in this chapter, a design goal of a long-distance fiber-optic system is to minimize the number of amplifiers required. To accomplish this goal, communications carriers use laser diodes that operate at wavelengths where attenuation is minimized, allowing the optical signal to be transmitted further before requiring the signal to be amplified. Figure 4.10 is a cross-sectional view of a typical laser diode.

The laser diode is manufactured using a process very similar to the production of large-scale integrated (LSI) and very-large-scale integrated (VLSI) semiconductor circuits. For this reason, the laser diode is also

Figure 4.10
A cross-sectional view of a laser diode.

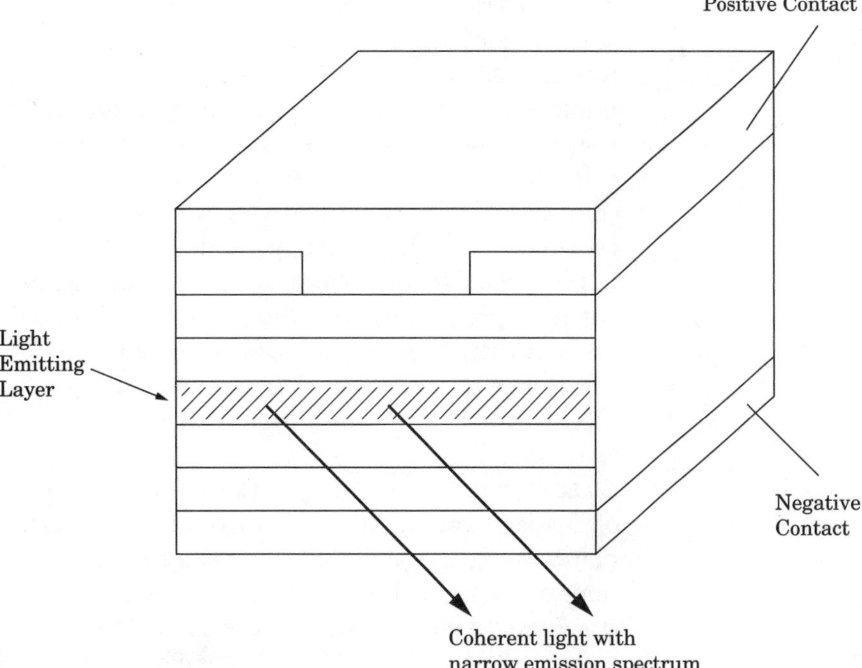

commonly referred to as a *semiconductor laser* and results in the fabrication of a laser diode that is relatively inexpensive to manufacture in quantity.

If you carefully examine Figure 4.10, you will note a series of layers that are employed to fabricate a laser diode. Those layers represent crystals that are grown as well as chemicals that are deposited on top of crystals. The top and bottom layers contain metal that functions as electrical contacts. As shown at the top of Figure 4.10, the top layer contains p-type regions that contain positive-current carriers. As noted earlier in this chapter, the p-type regions represent the absence of electrons or holes. In comparison, the n-type regions located at the bottom of the laser diode contain negative-current carriers. When an electrical current is applied to the contacts on the laser diode, electrons from the n-type layers flow into the p-type layers, where the electrons combine with the holes and the excess energy is emitted as coherent light.

If we again refer to Figure 4.10, we will note that the actual emission of light occurs in the central layer of the laser diode. The location of the emission is based on the structure of the layers surrounding the central layer. In other words, the adjacent layers are fabricated using material with a refractive index lower than that in the material in the central layer. In addition, the edges of the laser are fabricated to resemble mirrors, so that they reflect light back into the central layer. Those reflected photons stimulate the generation of additional light, which, in turn, results in the generation of an intense beam of coherent optical radiation. Because of the relatively narrow spectrum of light emission in comparison to an LED, it becomes possible and practical to couple the laser diode to thin-core, single-mode fiber.

Because of their higher power and quicker pulse rise time and coherent light generation, laser diodes are better suited for fiber-optic transmission that requires higher transmission rates over long distances. For this reason the use of laser diodes and single-mode fiber is the preferred mechanism for communications carriers.

SELECTION CRITERIA The selection of a laser diode depends on a number of factors, including the wavelength of the optical output and the output power, as well as the modulation rate of the device. In addition, the dimensions of the laser should generally be as small as possible. One parameter that is extremely important is the size of the emitter area; the smaller the emitter area, the easier it becomes to couple light into a single-mode fiber.

While most laser diodes have an emitting area between 1 and 3 μm that provides an efficient coupling into single-mode fiber, some lasers have much larger emitter areas. For example, very-high-power laser diodes may have an emitter area of approximately 100 μm. Such diodes can be efficiently coupled only to multimode fibers that have a similar core area. Thus, the type of fiber to be used governs the acceptable laser diode emitter area.

LASER DEVELOPMENTS Although lasers date to the 1960s, work continues at a fast pace to develop new types of devices more suitable for different applications, including optical communications. Two types of lasers that warrant attention are vertical-cavity surface-emitting lasers and tunable lasers.

VERTICAL-CAVITY SURFACE-EMITTING LASERS A *vertical-cavity laser* represents a laser formed by the vertical stacking of high-reflection mirrors. To obtain a high level of reflectivity, as many as 30 individual semiconductor layers are used to form each mirror. Between the mirrors there are several light-emitting semiconductor layers that result in the emission of light vertically upward from the surface of the semiconductor; hence the name *vertical-cavity laser.*

Vertical-cavity surface-emitting lasers (VCSELs) date to laboratory efforts during the mid-1980s. However, it wasn't until 1996 that Honeywell introduced a commercial VCSEL. The VCSEL emits light from the top surface of a semiconductor and can be mounted on top of a silicon photodiode integrated circuit, which reduces packaging cost. In comparison, this level of packaging was not possible with the use of commonly available edge-emitting lasers.

The first generation of VCSELs operate at 850 nm and are commonly used for gigabit transmission over multimode fiber at distances of a few hundred meters. Because transmission at longer distances is accomplished over single-mode fiber using 1310- or 1550-nm lasers operating in a transverse mode, which are several times more expensive than VCSELs, a significant amount of research is oriented toward the latter. More recent advances in VCSEL technology make it possible to manufacture this type of cost-effective laser in the 1300-nm operating range, and it is quite possible that 1500-nm VCSEL lasers may be available in the future.

VCSELs have several advantages over older semiconductor lasers, which emit light via their edges and are referred to as *edge-emitting devices.*

Advantages of VCSELs include lower production cost, lower packaging cost, increased reliability, lower power requirements, and simplified fiber coupling. The lower production cost results in the ability of VCSELs to be processed in batches of thousands of units at the same time. In comparison, edge-emitting lasers are created via a crystal-cleaving process that occurs either on an individual basis or on a row of devices. Although part of the production cost, packaging must also be considered. Edge-emitting lasers are hermetically sealed, which adds to their production cost. In VCSELs, however, the active regions are buried several micrometers beneath the VCSEL surface. This means that VCSELs have no exposed active parts and do not require hermetic packaging. In addition, this also means that VCSELs can be packaged in plastics instead of ceramics, which further reduces the manufacturing cost.

Because VCSELs have no exposed active parts, they have a higher level of reliability than do edge-emitting devices, where the active material is exposed. Because the VCSEL is considerably smaller than an edge emitter, it can support less costly light-coupling methods. In fact, connector coupling tolerances for VCSELs can be as liberal as 5 to 10 μm, compared to the much tighter 1-μm tolerance required by edge emitters.

Although VCSELs have many advantages, they are not problem-free. One key problem associated with the VCSEL is the fact that p-type material absorbs light much more strongly at 1300 nm than at 850 nm. As a result, it is difficult to design a VCSEL structure that has a sufficiently low level of absorption but is still laser-capable. Until the year 2000, this was the key problem preventing the development of long-wavelength VCSELs. However, during that year several manufacturers developed techniques that enable VCSELs to be fabricated with nonabsorbing mirrors. In addition, the use of different materials combined with nonabsorbing mirrors resulted in the creation of very efficient 1.3-nm VCSEL devices whose use in WAN communication systems is expected to represent a viable emerging technology.

TUNABLE LASERS A second generation of lasers that warrant discussion are tunable lasers. Although tunable lasers based on the use of a liquid medium have been available for many years, far more compact and less costly solid-state lasers were not available until quite recently. The development of tunable lasers is a necessity for implementing a lower-cost method of wavelength division multiplexing (WDM) that enables an array of wavelengths to be generated, each transporting a separate signal on a common fiber.

There are several types of tunable lasers, some of which are more suitable than others for optical communications. Organic-dye lasers use a liquid medium, in which the dyes are used to produce a range of vibration states that can generate a band of wavelengths from 300 to 1100 nm. Unfortunately, organic dyes degrade and result in the need for periodic replacement. A second more practical tunable laser is a vibronic laser. This type of tunable laser operates at a higher range of wavelengths than does a dye-based laser, and its solid-state composition makes it more practical for communications.

The basis for the operation of vibronic lasers is the use of certain solid-state media that allow a link between the applied electronic energy level and the vibration rate of the crystalline lattice of the medium. The use of several materials now permits tuning between 660 and 1180 nm.

The best type of tunable laser for communications is at present the tuning laser diode. Two different methods are used to tune this type of laser. In one method, which is referred to as a *distributed-feedback laser,* frequency is varied by changing either the temperature of the diode or the input current applied to the laser diode. This type of laser employs a grating structure fabricated along the active layer of the diode. The grating structure is spaced to provide feedback only for the wavelength specified by the grating spacing.

When an external-cavity laser is used as the foundation for a tunable laser diode, the cavity is employed as the tuning mechanism. Here the cavity is adjusted by placing grating beyond the active laser instead of along the active layer used by a distributed-feedback laser diode. Now that we have an appreciation for some of the developments occurring in the field of laser technology, let's turn our attention to another key component of an optical fiber transmission system. That component is the photodetector.

Photodetectors

While a light source is an important element of a fiber-optic transmission system, the photodetector is equally important. The photodetector senses light and converts it to electricity. Similar to the existence of multiple light sources, there are several types of photodetectors. However, before describing how they operate and discussing their advantages and disadvantages, let's consider their basic operation within a fiber-optic transmission system.

Basic Operation within an Optical System

Figure 4.11 illustrates the general relationship between a light source, an optical fiber, and a photodetector for a single-wave full-duplex transmission system. Note that two separate fibers are required in this operating environment, each with a separate light source and photodetector. The light source in the form of an LED or laser diode converts an electrical signal into an optical signal, while the photodetector performs a reverse process. Because each optical fiber represents a simplex or one-way transmission path, two optical fibers with attached light sources and photodetectors are required to support two-way, full-duplex transmission.

Figure 4.11 represents a basic optical transmission system. In this type of system light is modulated for transmission over the optical fiber and a photodetector is able to sense light only at the distant end. In a more complex optical transmission system, a light source can operate at two or more frequencies via the use of a tunable laser diode or multiple lasers operating at different frequencies that are coupled onto a common fiber. In this situation the photodetector must be capable of detecting light at the tunable frequencies generated by the laser diode or the frequencies generated by multiple lasers. This type of transmission system supports multiple derived channels via different optical frequencies and is referred to as *wavelength division multiplexing* (WDM). When the number of derived channels is substantial, the multiplexing technique is referred to as *dense wavelength division multiplexing* (DWDM), and the photodetector must respond to a wider range of frequencies.

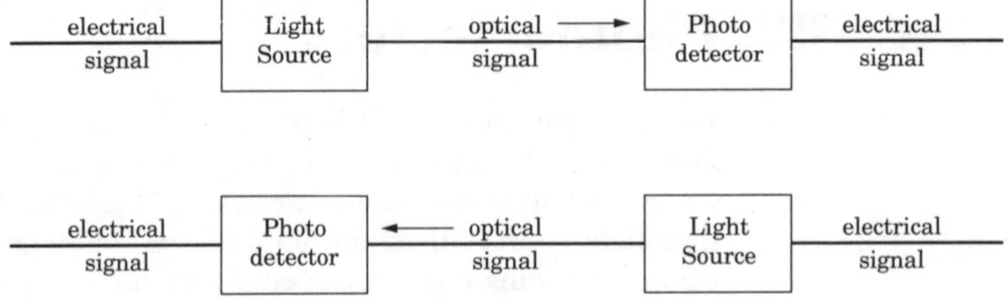

Figure 4.11 The relationship between a light source and an optical detector.

Radiation Absorption

A *photodetector* is an optical detector that absorbs radiation and outputs an electrical signal that is normally proportional to the intensity of the detected electromagnetic radiation. As a refresher from Chapter 2, the energy of a photon is given by the formula

$$E = \frac{hc}{\lambda}$$

where h = Planck's constant [6.626×10^{-34} J/s (joules per second)]
c = speed of light (2.9979×10^8 m/s) , which is normally rounded to 3×10^8 m/s in most publications
λ = wavelength

The number of photons per unit area of a detector striking it per second can be expressed as $N = H\lambda/hc$, where H is the intensity of the incident radiation in units of power per unit area.

From the preceding observations and our knowledge of atoms and electrons described earlier in this chapter, we know that electrons can be excited to the conduction band through the absorption of photons. Because the bandgap between the valence and conduction bands in an atom is well defined, it becomes possible to create a semiconductor photodetector whose atoms are excited to the conduction band during photon bombardment. Because photons with an energy equal to or greater than the bandgap energy level can excite valence electrons while those with a lesser amount of energy cannot, this means that only radiation with wavelengths shorter than a critical value is capable of generating conduction electrons. This results in a critical wavelength that can be expressed in terms of the energy across the bandgap E_g as $\lambda = hc/E_g$.

This means that it is possible to detect radiation at a specific wavelength by using semiconductor material that is fabricated with a given bandgap, and this provides the basis for the creation of photodetectors. The amount of energy required for a particular bandgap varies from material to material. For visible radiation, silicon is employed as it can detect radiation with wavelengths under 1 μm. However, unless silicon is doped with specific impurities to define specific bandgap energy levels, it is incapable of detecting infrared radiation.

INFRARED REGIONS As a refresher, visible light extends from wavelengths of approximately 400 to 700 nm; wavelengths shorter than visible light are referred to as *ultraviolet, X rays,* and *gamma rays,* as the wavelength decreases. The region from 700 nm to 100 μm is referred to as the *infrared region,* with wavelengths from 700 nm to 1 μm considered to represent the near-infrared region. The remainder of the infrared region is additionally subdivided into the shortwave infrared region (SWIR) from 1 to 3 μm, the midwave infrared region (MWIR) from 3 to 5 μm, and the longwave infrared region (LWIR), whose wavelength exceeds 8 μm. As a result of strong atmospheric absorption, the region between 5 and 8 μm is commonly ignored.

DOPING MATERIAL The use of lead sulfide within a semiconductor provides a mechanism to detect radiation under 3 μm. Thus, it provides a basis for developing a SWIR detector. Other frequently used material for semiconductor detectors include silicon carbide, lead selenide, and indium antimonide. Silicon carbide results in the creation of a relatively wide bandgap, which results in the ability to detect ultraviolet radiation. The use of lead selenide and indium antimonide results in the ability to detect radiation at a wavelength under 5 μm and provides the foundation for an MWIR detector. Thus, the doping of impurities provides the basis for the creation of detectors literally "tuned" to different wavelengths.

Types of Photodetectors

There are two general types of photodetectors: (1) those that are semiconductor-based and have a bandgap that can be broken through by incident radiation and (2) non-semiconductor-based detectors that do not employ a bandgap bridge. Semiconductor-based photodetectors are used primarily in optical communications systems and are also commonly referred to as *quantum detectors.*

DETECTOR METRICS There are two basic types of semiconductor based photodetectors, photoconductors and photodiodes, both of which are commonly characterized by the same general metrics. Those metrics include their responsiveness, noise level, response time, and detection sensitivity. Since these metrics are the same for photoconductors and photodiodes, let's identify them before examining each type of detector.

RESPONSIVENESS The *responsiveness* of a detector represents the ratio of output in terms of voltage or current to the input in terms of the optical power incident on the detector. Responsiveness is normally expressed in terms of volts per watt (V/W) or amperes per watt (A/W). As you might expect, a higher level of responsiveness is preferred.

NOISE There are two types of noise associated with a photodetector: white noise and 1/*f* noise. *White noise* represents background noise generated by the movement of electrons and has a uniform spectral power density. Some detectors have an additional noise component referred to as 1/*f noise*, whose spectral power density is inversely proportional to frequency. Because a higher signal-to-noise ratio (S/N) is more advantageous, it is important to obtain the former. To do so requires a reduction in bandwidth that is normally accomplished by the use of filters. Because 1/*f* noise decreases with increasing frequency, this permits the S/N ratio to be maximized via light detection at a frequency sufficiently high that 1/*f* noise can be minimized.

RESPONSE TIME The *response time* represents the time required for a detector to respond to a significant change in input optical power. A lower response time enables the detector to operate at a higher data transmission rate. Now that we have an appreciation for the basic metrics associated with the performance level of a photodetector, let's discuss each type of detector.

PHOTOCONDUCTORS A *photoconductor* represents a type of semiconductor that has heavily doped n-type or p-type material. This type of photodetector is commonly used for infrared detection.

During operation, a photoconductor's electrical conductivity increases linearly with temperature. Thus, the photoconductor provides an increase in current when the detector is exposed to radiation. In contrast, the absence of radiation produces a current referred to as the "dark current."

PHOTODIODES A diode represents a region of a single-crystal semiconductor material that has a p-n junction, resulting in electrons in the conduction band on the n side of the junction drifting across to combine with the holes on the p side of the junction. After a period of time the n-type material near the junction is depleted of conduction electrons while the p-type material near the junction is depleted of holes. This results in the generation of a potential difference across the p-n junction.

When a photodiode receives optical radiation with a wavelength short enough to bridge the bandgap at the p-n junction, additional electrons from the n region flow across the p-n junction and combine with holes in the p region. This results in a widening of the depletion region and causes the potential across the p-n junction to widen until electrostatic forces again act to stop the flow of electrons. When the gap widens, current ceases to flow through the diodes, resulting in the generation of a voltage that is proportional to the intensity of the received optical radiation. This action results in an open circuit; thus, the voltage across the p-n junction is called the *open-circuit voltage* V_{oc}.

A photodiode can also operate in a short-circuit condition by short-circuiting both sides of the detector to prevent voltage from occurring across the diode. This action results in a current flowing through the diode, referred to as the *short-circuit current* I_{sc}. Here I_{sc} is proportional to the incident intensity and provides a second mechanism for converting an optical signal into an electrical signal.

Since 1990 or so, various materials have been used for infrared photodiodes, some of which were mentioned earlier in this chapter. While a complete list of photodiode material is beyond the scope of this book, it should be noted that manufacturers developed products that operate at speeds as high as 100 GHz by the mid-1990s. It should also be noted that the selection of an applicable light detector depends on several factors. Those factors can include the operational wavelength, speed of detection (which is a function of response time), S/N ratio, and responsiveness. Another important consideration when selecting a high-speed detector is the difference between time-domain and frequency-domain optimization. A photodetector optimized for frequency-domain applications exhibits a frequency response that is relatively flat over the frequency band, while the time domain is relatively wide. In comparison, a photodetector optimized for time-domain applications has a narrower detector pulse width value but its frequency response falls off with increased frequency. Figure 4.12 provides a general comparison of photodetectors optimized for frequency-domain and time-domain applications.

Couplers and Connectors

Our discussion of light sources, detectors, and fibers would not be complete without a brief discussion of couplers and connectors.

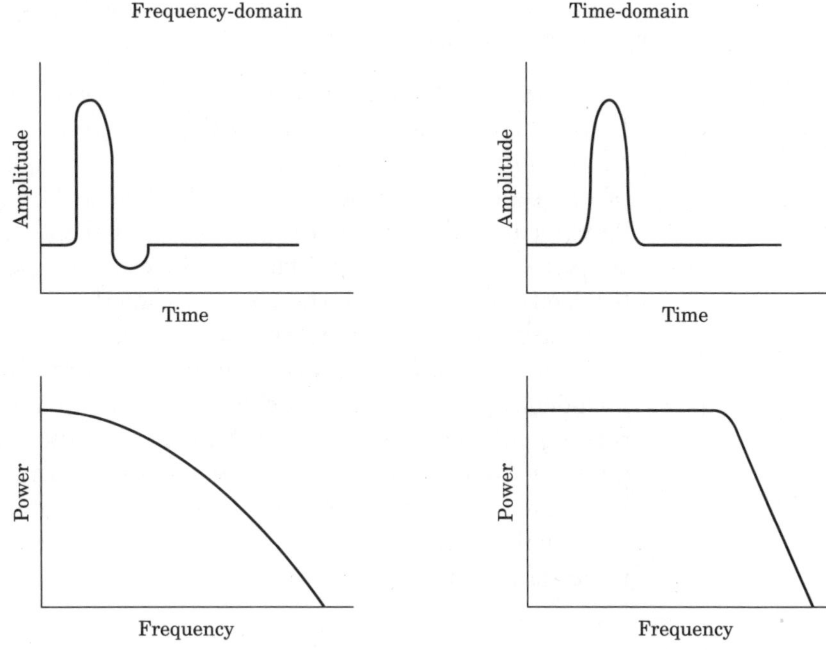

Figure 4.12
A general comparison of the operation of photodetectors optimized for frequency-domain and time-domain applications.

Function

The job of the coupler is to couple light from an optical source into a fiber or from a fiber into an optical detector. This deceptively easy task is performed in three key steps:

1. The divergent light from LEDs and to a lesser extent from lasers must be converted into a beam of parallel light. This process, which is referred to as *collimation,* is performed by placing the end of a fiber at the focal plane of a lens.
2. The collimated beams must be focused into a location or spot that matches the properties of the optical fiber, which is the core of the fiber. This process can be considered as collimating in reverse and requires the spot diameter from the focused beam to be less than or equal to the mode field diameter (MFD) of the optical fiber. Here the mode field diameter is a measure of the intensity of light traveling within a fiber where the intensity drops to 13.5 percent of the peak intensity. For a single-mode fiber the MFD is approximately 15 percent greater than the actual size of the core.

3. Precise alignment of optics is necessary to couple light from one medium to another.

Figure 4.13 illustrates the relationship between a laser mounted in a housing, a collimating lens, a focusing lens, and an output fiber. As illustrated in this figure, the entire apparatus can be considered representative of one type of fiber coupler. Actually, there are several types of couplers developed by different manufacturers, and the selection of a particular coupler depends on several factors. These factors include the diameter of the laser diode housing, the connector method, the coupling efficiency, and the diameter of the fiber to be coupled. Concerning the connector method, various products on the market when this book was written included support for screw locking, pigtail coupling, and threading. Concerning coupler efficiency, this is normally closely coupled to the laser diode or fiber and converts it into a beam of parallel light. Although a laser's light is less diverse than that of an LED, a collimator is required for both light sources. If we focus on the collimator lens, we can begin to understand how the collimating process operates, so let's do so.

The Collimating Process

Figure 4.14 illustrates the operation of a fiber collimator. By extrapolation from optical relationships, it is possible to predict the behavior of the collimated beam. For example, if the focal length of the lens is *f*,

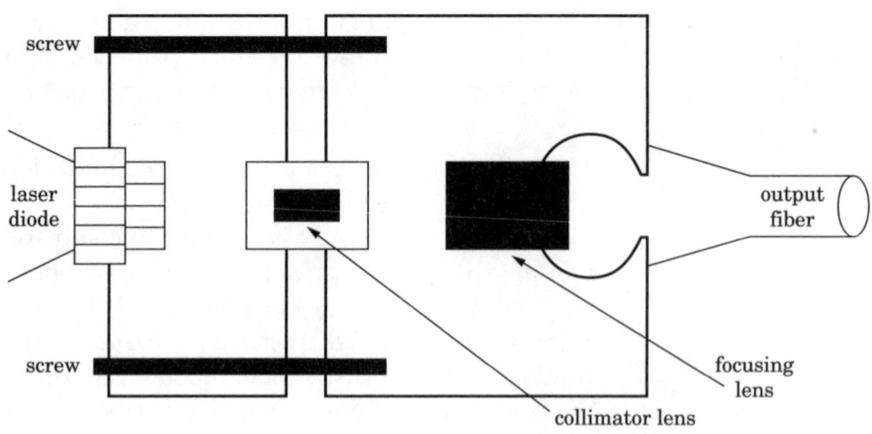

Figure 4.13 There are many types of laser diode to fiber-optic couplers with differences in the housing, lenses, and coupling method supported. The coupler shown uses both a collimating lens and a focusing lens with locking screws to fasten the coupler together.

then the characteristics of the collimated beam are given by the following formulas:

$$BD = 2 \times f \times NA$$

$$DA = \frac{MFD}{f}$$

where MFD and NA are the mode field diameter and numerical aperture of the waveguide while BD and DA represent the beam diameter and full-divergence angle of the collimated beam. The BD and focal length of the lens are expressed in millimeters. In comparison, the DA is expressed in milliradians, where 1 mrad is 0.057° while the MFD is expressed in micrometers.

As shown in Figure 4.14, large waveguides do not collimate well. For example, suppose you attempt to collimate a 500-μm-core fiber with a 5-mm-focal-length lens. In this situation the divergence angle becomes 500 μm/5 mm or 100 mrad. Since 1 mrad is 0.057°, the DA becomes 5.7°, which may not be much better than what you started with prior to the collimating process.

Mode Field Diameter and Numerical Aperture Considerations

A second point that warrants attention is how the NA and MFD are defined. Laser manufactures define their beams in terms of gaussian

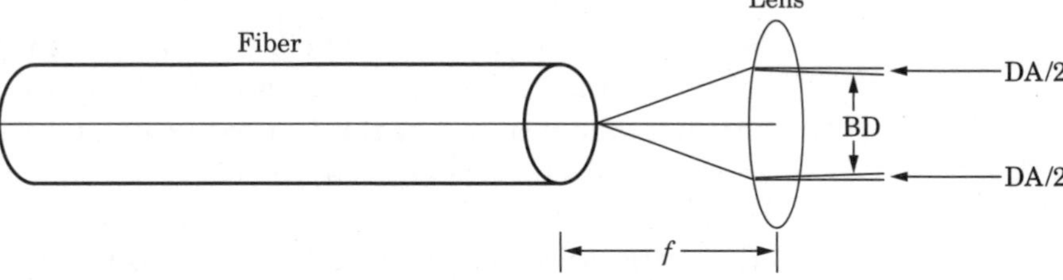

Beam Diameter (BD) = $2 \times f \times NA$
Divergence Angle (DA) = MFD/f

f = focal length of the lens
NA = Numerical Aperature of the waveguide
MFD = Mode Field Diameter

Figure 4.14 The operation of a fiber collimator.

beam characteristics. As a result, the beam size of a laser diode is defined at the point where its intensity drops to 13.5 percent of its peak intensity of $1/e^2$. Laser diode manufacturers refer to this as *full-width at half-maximum* (FWHM) levels. If you are coupling a laser diode to multimode fiber, you can use the core diameter as the MFD and the refractive indices of the fiber's core and cladding to compute the fiber's NA. To do so, you would use the following formula:

$$\text{NA} = \sqrt{N_1^2 - N_2^2}$$

where N_1 is the index of refraction of the core of the fiber and N_2 is the index of refraction of the cladding of the fiber.

Although the preceding formula is acceptable for multimode fiber, the use of single-mode fiber can result in significant computation errors. A more accurate method when working with single-mode fiber is to treat the output of the fiber as a gaussian source. Then the NA can be computed as NA = 2l/MFD, where l is the wavelength of light expressed in terms of micrometers. According to the optics literature, the resulting NA value computed for single-mode fiber can be up to a third smaller than the value listed by some fiber manufacturers.

When selecting a lens, a good rule of thumb to follow is to select optics with an NA at least 50 percent greater than the waveguide to be collimated. Doing so can result in over 99 percent of transmission flowing through the optics.

Coupling Considerations

Once a light beam is collimated, you need to consider the coupling of the beam into a fiber or waveguide. In doing so you need to consider the spot diameter (SD) of the focused beam and its NA. The focused SD should be less than or equal to the MFD of the waveguide, while the NA of the focused beam should not exceed the NA of the waveguide. If you know the focal length f of the lens, you can compute the SD and numerical aperture for the focused beam NA_B as follows:

$$\text{SD} = f \times \text{DA} \times \text{MFD}$$

$$\text{NA}_B = \frac{\text{BD}}{2} \times f \times \text{NA}_{\text{fiber}}$$

If you are using multimode fiber, you should select a lens that provides values for SD and NA_B less than or equal to 70 percent of the maximum allowable values. In addition, you should not attempt to focus the spot on an area less than 30 percent of the core of the fiber, as this could cause burning of the end of the fiber. While it is possible to find a range of focal length lenses that satisfy the previously listed computations, when working with single-mode fiber, only one focal length will satisfy spot size and numerical aperture requirements.

Alignment Considerations

Once you select collimating and focusing optics, you need to align the optics to minimize loss. To do so, you should follow a standard industry technique, which is to first focus light on a spot similar to the size of the waveguide MFD. Once the preceding is accomplished, you would position the waveguide until its endface is located at the focal point. Because MFDs of communications systems are approximately 10 μm or less, it is very difficult to perform the positioning mentioned above. This is because positioning equipment that supports submicrometer resolution along all three axes (*x*, *y*, and *z*) necessary to align a waveguide is bulky, expensive, and difficult to use. As an alternative, you can first attach a lens to the waveguide to generate a collimated beam. This enables you to work with collimated optics and obviates the need to consider the distance between the waveguide and the bulk optics. This technique also reduces the precision required for lateral alignment; however, similar to the fact that there is no free lunch, working with a collimated beam is not problem-free. The key problem is the fact that sensitivity to angular misalignment will now increase. One method for correcting this problem involves the use of a tilt-adjustable coupler. This type of coupler uses screws to adjust the tilt angle between the focusing optics and the collimated beam. This enables the use of the screws to adjust the tilt angle, which in turn permits adjustment of the focused rays to strike the core of the fiber.

Types of Couplers and Connectors

There are numerous types of couplers. Some couplers couple light from a collimated source into either single-mode, multimode, or polarization-maintaining fibers. Couplers are available for different wavelengths and for different input powers. In addition, couplers vary according to the light source as well as the connector. Concerning the latter, common connectors include a receptacle-style physical contact, receptacle-style noncontact style, and pigtail-style couplers. Each coupler has advantages and disadvantages associated with its use that are application-dependent. These features should be carefully considered if you are designing a system instead of purchasing an electricooptical system, as the latter includes light source, coupler, connector, and fiber as a single entity.

CHAPTER 5

Fiber in the LAN

In this chapter we discuss the use of transmission over fiber-optic media in the local area network (LAN) environment. Because the use of optical fiber in a LAN dates to the early 1980s with the introduction of the Fiber Distributed Data Interface (FDDI) standard, we will begin our examination of fiber in the LAN with a brief examination of FDDI. Although FDDI is relatively expensive and is considered by many persons to be obsolete because of the low availability of low-cost Fast Ethernet and the reduced cost of Gigabit Ethernet, to paraphrase Mark Twain, rumors of its death have been greatly exaggerated. FDDI uses a token-passing scheme which provides a data transfer capability that approaches 90 percent of its 100-Mbits/s operating rate, while lower-cost Fast Ethernet in a shared media environment will have collisions because of its design, which reduces practical throughput considerably under 100 Mbits/s. Once we examine FDDI, we will turn our attention to Ethernet and Fast Ethernet, examining the use of fiber in both network environments. We will then conclude this chapter with a discussion of Gigabit Ethernet.

Fiber Distributed Data Interface (FDDI)

The FDDI standard dates to October 1982, when the American National Standards Institute (ANSI) Committee X3T9.5 was chartered to develop a high-speed data networking standard. The resulting effort was the development of a token-passing optical fiber media network based on the use of two fiber pairs, each operating at 100 Mbits/s.

Position in the OSI Reference Model

FDDI is defined as the two bottom layers of the seven-layer International Systems Organization (ISO) Open System Interconnection (OSI) Reference Model: the physical and data-link layers. Figure 5.1 illustrates the relationship of FDDI, Ethernet, and Token Ring LANs to the seven-layer ISO OSI Reference Model. In examining Fig. 5.1, note that FDDI is similar to Ethernet and Token Ring in that it provides a transport facility for higher-level protocols such as TCP/IP. In fact, FDDI was, and in some locations still is, commonly used at the Internet Service Provider (ISP) peering points that provide interconnections between ISPs.

The FDDI standard recognized the need to subdivide the physical layer. Under the FDDI standard the physical-medium-dependent (PMD) sublayer defines the details of the fiber-optic cable used, while the physical (PHY) layer specifies encoding/decoding and clocking operations. Figure 5.2 illustrates the FDDI subdivision of the physical layer. Later LAN standards, such as Fast Ethernet and Gigabit Ethernet, can be considered to use the learning curve from FDDI as they also resulted in the subdivision of the physical layer.

4B/5B Coding

The actual coding method used by FDDI results in the encoding of 4 bits with the use of a 5-bit pattern. Thus, this encoding technique is referred to as a 4B/5B code. Because 4 bits are encoded into 5, this means there are 16 4-bit patterns. Those patterns, which are listed in Table 5.1, were selected to ensure that a transition is present at least twice for each 5-bit code. Because 5-bit codes are used, the remaining symbols provide special meanings or represent invalid symbols. Concerning special

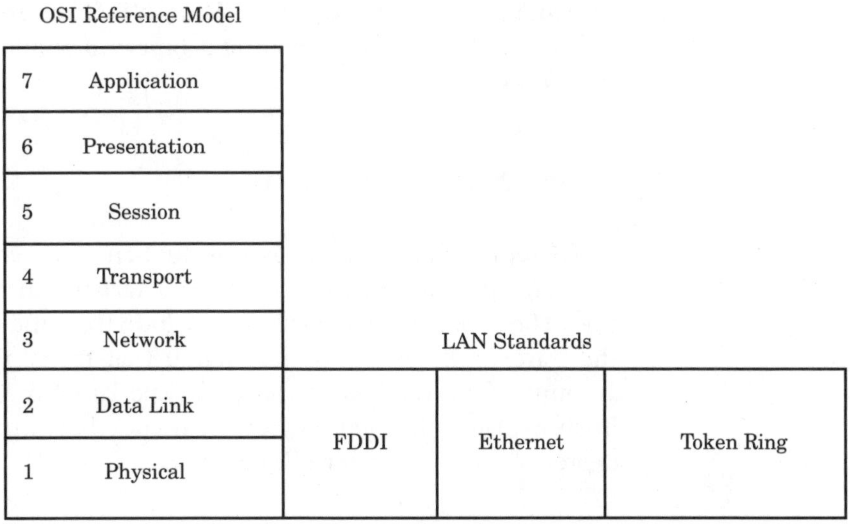

OSI Reference Model

7 Application

6 Presentation

5 Session

4 Transport

3 Network

2 Data Link

1 Physical

LAN Standards

FDDI

Ethernet

Token Ring

Figure 5.1 The relationship between the ISO OSI Reference Model and common LAN standards.

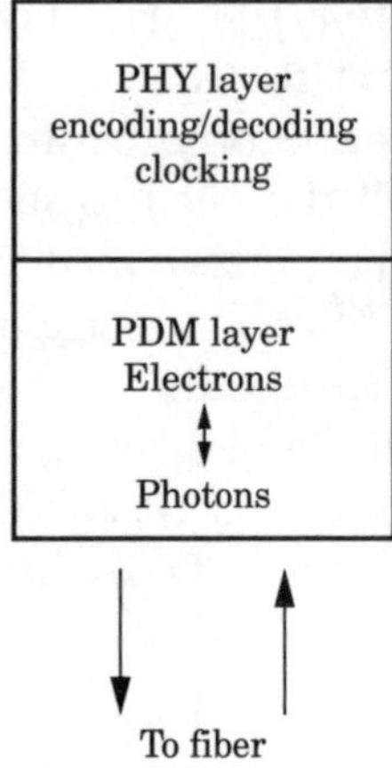

Figure 5.2
The FDDI standard subdivides the physical layer.

meanings, the I symbol is used to exchange handshaking between neighboring stations, while J and K symbols are used to form the Start Delimiter for a packet, which functions as an alert to a receiver that a packet is arriving.

The selection of the 4B/5B coding was based on the need to reduce the signaling level from 200 MHz to a 125-MHz rate, which, when FDDI was standardized, considerably reduced the cost of electronic components. FDDI uses 4B/5B encoding to operate on 4 bits at a time, converting them into a 5-bit code. Then each bit is encoded using non-return-to-zero-inversion (NRZI) transmission. As a refresher, NRZ codes a 0 as zero and a 1 bit as a plus pulse. Thus, NRZI reverses that coding, resulting in a pulse of light signaling a binary 0, while the absence of light signifies a binary 1.

Fiber Specifications

FDDI represents a token-passing dual-ring network. The ANSI PMD standard specifies several optical characteristics that govern FDDI operation. Those optical characteristics include the type of optical fiber used, the wavelength of the optical signal that transmits the data, and the amount of power loss in the cable; the latter is specified in terms of both a power budget expressed in decibels and cable attenuation expressed in decibels per kilometer.

TABLE 5.1

FDDI 4B/5B Codes

4-bit code function	5B code	Symbol
Starting Delimiter		
First symbol	11000	J
Second symbol	10001	K
Data symbols		
0000	11110	0
0001	01001	1
0010	10100	2
0011	10101	3
0100	01010	4
0101	01011	5
0110	01110	6
0111	01111	7
1000	10010	8
1001	10011	9
1010	10110	A
1011	10111	B
1100	11010	C
1101	11011	D
1110	11100	E
1111	11101	F
Ending Delimiter	01101	T
Control indicators		
Logical 0 (reset)	00111	R
Logical 1 (set)	11001	S
Line status symbols		
Quiet	00000	Q
Idle	11111	I
Halt	00100	H

TABLE 5.1

FDDI 4B/5B Codes (Continued)

4-bit code function	5B code	Symbol
Invalid-code assignments		
	00001	Void or halt
	00010	Void or halt
	00011	Void
	00101	Void
	00110	Void
	01000	Void or halt
	01100	Void
	10000	Void or halt

OPTICAL FIBER SUPPORT FDDI can support 62.5/125-, 50/125-, and 100/140-μm multimode fiber sizes; 62.5/125 multimode fiber is the preferred medium. FDDI also supports the use of single-mode fiber, which is commonly used for long-distance transmission that enables the technology to extend to a metropolitan area network (MAN) of up to 100 km. When single-mode fiber is used, it has a core diameter of 8 to 10 μm and a cladding diameter of 125 μm. Thus, FDDI single-mode fiber is commonly specified as 8/125, 9/125, and 10/125.

OPTICAL TRANSMITTER Available transmitters and receivers operate at 850, 1300, and 1550 nm; the 1300 nm wavelength is most commonly used. Transmission at 850 and 1300 nm is designed for use over multimode fiber. Single-mode fiber that can support FDDI for greater transmission distances operate at 1300 and 1500 nm. Both lasers and laser diodes are used with multimode fiber; however, when single-mode fiber is used, the transmitter must be a laser diode since it has the narrow wavelength required to transmit optical signals into the narrow core of the fiber.

ATTENUATION For multimode fiber the FDDI PMD standard specifies a power budget of 11.0 dB. Because the power budget of a system represents the minimum transmitter power and the minimum receiver sensitivity, this means that up to 11 dB of the optical signal can be lost.

In addition, the maximum cable attenuation is 1.5 dB/km at 1300 nm. When single-mode fiber is used, the PMD specifies a range of power budgets. The specified range depends on the types of transmitters used and extends from 10 to 32 dB (maximum).

Media Interface and ST Connectors

One of the well-thought-out portions of the ANSI standard is represented by the FDDI *Media Interface Connector* (MIC). This connector, which is used to connect multimode fiber to an FDDI station, consists of a keyed plug and a keyed receptacle and is polarized to ensure that an appropriate transmitter/receiver-to-fiber association occurs, representing perhaps the first "dummyproof" optical connection.

FDDI also supports ST-type connectors to physically connect optical fiber to an FDDI station. Although ST connectors provide a lower-cost alternative for connecting optical fiber to an FDDI station, they do not provide a polarized receptacle. Thus, when using an ST-type connector, you must ensure that transmit and receive connections are not reversed.

Ring Structure

Figure 5.3 illustrates how FDDI can compensate for a cable break. The left portion of the figure illustrates the use of the dual-ring structure to interconnect four workstations. Note that the FDDI backbone consists of two separate fiber-optic rings, one of which is active and referred to as the *primary ring*, while the other ring can be considered to be "on hold," waiting to be activated, and is referred to as the *secondary ring*. Each device directly connected to the FDDI backbone must be attached to both the primary and secondary rings, resulting in the term *dual attachment*. If a problem occurs on the primary ring, the secondary ring will be used. In fact, FDDI's Station Management (SMT) facility built into each FDDI adapter on either side of a cable break or another impairment on a cable recognizes that communications have failed on the ring. SMT then wraps the communications from the primary ring back out onto the secondary ring in the opposite direction, permitting all stations other than those in that portion of the ring that is broken to resume communications. The middle portion of Figure 5.3 illustrates the occurrence of a break in the FDDI network, while the right portion of that illustration shows how SMT interconnects or wraps communications from the primary ring onto the secondary ring in the opposite direction to re-form the ring. Thus, a break in an FDDI ring can be considered as similar to the battery commercial; that is, like the bunny in the commercial, the FDDI ring keeps on operating.

Station Types

Two station types are defined under the FDDI standard: class A and class B. Class A stations connect to both primary and secondary rings that make up an FDDI network and are better known as *dual-attachment stations.* Class B devices are indirectly connected to a ring via a class A device. Thus, a class A station acts as a wiring concentrator that connects several class B stations to the FDDI ring. As you might expect, a class B station is referred to as a *single-attachment station.*

Figure 5.4 illustrates the use of several single-attachment stations and one dual-attachment station on an FDDI ring. Note that this configuration is designed as an economy measure since it would have been too expensive during the 1980s, when FDDI was developed with SMT capability for each device connected to the LAN. Thus, restricting SMT to dual-attachment stations results in the wrapping of communications during a cable break or another impairment supporting redundancy for

Figure 5.3 Compensating for a cable break.

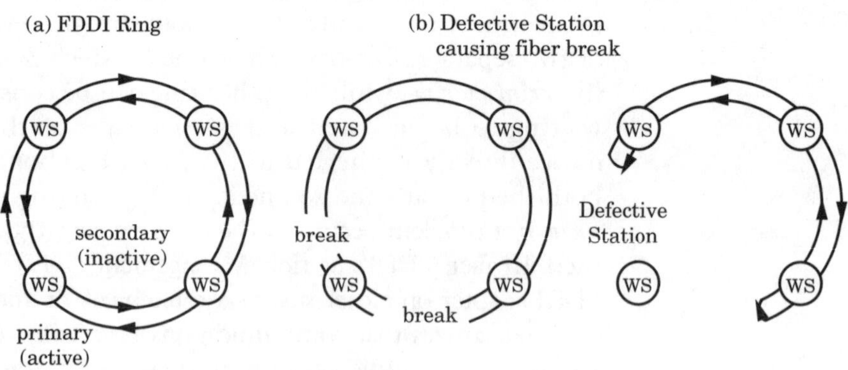

Legend:  workstation

multiple stations. Because of the high degree of redundancy afforded through the use of FDDI, it remains a popular infrastructure at Internet peering points.

Port Types and Rules

To prevent the incorrect configuration of a network, ANSI FDDI standards specify certain connection rules. Those rules cover the functions of four defined port types: A, B, M, and S. Those ports govern connectivity between workstations and a concentrator as well as the connection of a concentrator to the primary and secondary rings. Figure 5.5 is a schematic diagram of the configuration of these ports in an FDDI network.

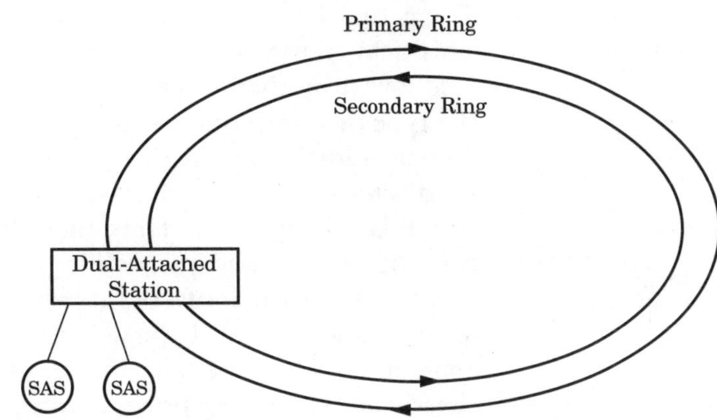

Figure 5.4
A dual-attachment station acts as a connection for the indirect attachment of single-attachment stations.

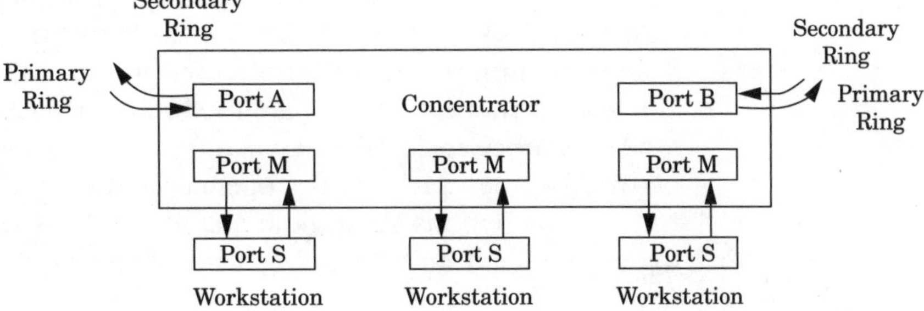

Figure 5.5
FDDI port types.

In examining Figure 5.5, it is important to note that each port connector has a specific mechanical keying interface defined by the PMD standard. For example, port A connects to the incoming primary ring and outgoing secondary ring. In comparison, port B connects to the outgoing primary ring and the incoming secondary ring. In comparison, port M provides connectivity to a single-attachment station or to another concentrator, while port S connects a single-attachment station or single-attachment concentrator.

FDDI Frame Format

The basic FDDI frame format is illustrated in Figure 5.6. The preamble is variable in length, consisting of a minimum of 16 4B/5B I symbols. The actual beginning of the frame is the Start Delimiter (SD) field. This field consists of the 4B/5B J and K control symbols. The SD field is followed by the Frame Control (FC) field. This 8-bit field identifies the type of frame and is followed by the Destination Address (DA) and Source Address (SA) fields. Each field is 48 bits in length and corresponds to the MAC (Media Access Control) addresses used by Ethernet and Token Ring LANs. Thus, the first 3 bytes represent the manufacturer of the FDDI adapter, while the last 3 bytes represent the specific adapter number manufactured by the vendor. The source address is followed by a variable Information field and a 4-byte Frame Check Sequence (FCS) field. The FCS contains a 32-bit cyclic redundancy check (CRC) value that provides integrity for the FC, DA, SA, and information fields. The End Delimiter (ED) field consists of two 4B/5B T symbols.

FRAME TYPES FDDI specifies several types of frames, each of which has a defined function. Those frame types include MAC frames, SMT frames, and LLC frames. MAC frames carry *Media Access Control* data. Such frames include claim frames used in the ring initialization process and beacon frames; the latter are used in the ring fault-isolation process.

SMT frames are used to transport FDDI management information between frames. Such frames operate, control, and maintain the FDDI ring and its stations. As indicated earlier, SMT is responsible for detecting a cable fault and wrapping a primary ring to a secondary ring to reroute data.

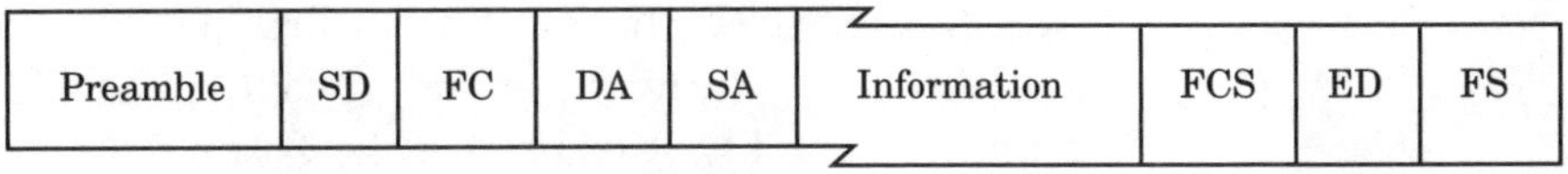

Preamble	SD	FC	DA	SA	Information	FCS	ED	FS

Figure 5.6
The FDDI information frame.

Legend:

Preamble	16 or more I symbols
SD	Start Delimiter
FC	Frame Control
DA	Destination Address
SA	Source Address
FCS	Frame Check Sequence
ED	End Delimiter T symbol
FS	Frame Status

A third type of frame transported via FDDI is the *Logical Link Control* (LLC) frame. LLC frames transport information on the FDDI network. Whereas MAC and SMT frames are local to a ring, LLC frames can be bridged or routed off a ring.

RING SCHEDULING AND SERVICES FDDI supports the operation of a *timed-token protocol* (TTP), which defines the means by which a station acquires access to a ring. As we will soon note, several timers govern the operation of an FDDI LAN. However, before discussing the role of timers, we shall describe the two types of services supported by FDDI: asynchronous and synchronous.

ASYNCHRONOUS SERVICES In FDDI terminology, *asynchronous transmission* refers to the ability of a station to transmit data during periods when bandwidth is not reserved by synchronous services. Thus, we can note that in the FDDI context the term *asynchronous services* represents the transmission of information that is not extremely delay-sensitive (i.e., not time-critical).

SYNCHRONOUS SERVICES A second type of service specified by FDDI is synchronous transmission services. In *synchronous services*, each station is guaranteed a portion or slice of the 100-Mbit/s FDDI bandwidth. The actual amount of synchronous bandwidth allocated to a station is negotiated using an allocation procedure defined by SMT. As we will soon note, when a station does not need its guaranteed bandwidth for synchronous services, it can use that bandwidth for asynchronous services.

TOKEN PASSING If you are familiar with the 4- or 16-Mbit/s Token Ring LAN frame format, you are probably mystified as to the location of the token in the FDDI frame. Unlike Token Ring, which includes a token bit in its Frame Control field, FDDI uses a special six-symbol frame as illustrated in Figure 5.7. The token frame consists of four fields: a preamble field of 16 or more I symbols, an SD field of J and K symbols, an FC field, and an ED field. Stations that have data to transmit must first acquire a token. However, unlike a Token Ring LAN where the token is part of the frame and is transmitted within the formed frame, under FDDI the token is held by the transmitting station, a process referred to as *token absorption.* The actual period of time that the token is held is specified by a *timed-token protocol* (TTP). A station can transmit as much data as it has during its token hold time or until the hold time expires. Actually, each station on an FDDI ring uses three timers to regulate its operation. These timers are described in the following three paragraphs.

TOKEN ROTATION TIMER The *token rotation timer* (TRT) is used to time the duration of operations permitted by a station. The value assigned to the TRT depends on the state of the ring. For instance, during steady-state operations the TRT will expire when the target token rotation time (TTRT) is exceeded. Stations negotiate the value for the TTRT during the claim process, which determines the station that will initialize the FDDI ring. This process represents an auction in reverse, where the station with the lowest bid for the TTRT time acquires the right to initialize the ring.

TOKEN HOLDING TIMER The *token holding timer* (THT) controls the period of time during which a station can operate asynchronously. FDDI supports both asynchronous and synchronous services; the former designed to support bandwidth-intensive operations when the station does not need the guaranteed bandwidth provided by synchro-

Figure 5.7
FDDI token frame.

Preamble	SD	FC	ED

Legend:
SD Start Delimiter (2 symbols)
FC Frame Control (2 symbols)
ED End Delimiter (2 symbols)

nous services. Thus, asynchronous frames are transmitted only when bandwidth reserved for synchronous operations is not used. In comparison, synchronous frames can be transmitted at any time as long as the negotiated synchronous bandwidth allocation is not exceeded. The token holding time for a station is initialized with the value corresponding to the difference between the arrival of the token and the TTRT value. To transmit asynchronous frames, a station must have a token and the THT timer must not have expired.

VALID TRANSMISSION TIMER A third FDDI station timer is the valid transmission timer (TVX). The function of this timer is to time the period between valid transmissions on the FDDI ring. If a token is lost, excessive noise occurs on the ring, or another impairment results in the TVX expiring, the station assumes a problem occurred and initiates a ring initialization operation to restore the ring to an operational status.

Summary

FDDI represents a well-thought-out but complex fiber-optic LAN technology. While it can support a dual ring up to 100 km (62 miles), which makes it attractive for MAN operations, its primary use is in a LAN environment. Unfortunately, its relatively high cost because of its complexity has limited its appeal in comparison to the Fast Ethernet and Gigabit Ethernet, which benefit from lesser complexity and greater economies of scale; the latter advantage is due to the considerably larger market for the Ethernet products. Nevertheless, FDDI continues to be used in mission-critical areas where reliability as well as the ability to transmit data near the maximum LAN operating rate is critical.

Ethernet and Fast Ethernet

Anyone who has not been in a prolonged period of isolation would know that the battle for LAN dominance is over, with Ethernet in its several flavors representing the victor. However, what we may not realize is the fact that fiber-optic components have a long association with Ethernet, dating from their use as a transmission extender for 10 Mbit/s

Ethernet to the preferred media used to construct a Gigabit Ethernet network. In this section we will focus on the use of optical components to extend the capabilities of different types of Ethernet networks. However, because fiber-optic media are used primarily to extend transmission distance in an Ethernet and Fast Ethernet environment, we will begin our examination of the use of fiber by focusing on the network diameter of an Ethernet LAN.

Network Diameter Constraints

Ethernet uses a *Carrier Sense Multiple Access with Collision Detection* (CSMA/CD) access protocol. Under this protocol a station with data to transmit first listens to determine if another station is transmitting or if the network is idle. If the network is idle, the station will transmit a frame; otherwise, the station waits for an idle condition. Once the station begins transmission, it will listen for a short period of time to determine if a collision occurred. That period of time is referred to as the Ethernet *slot time* T_s and is defined as the time duration required for the transmission of 512 bits of data (64 bytes). The selection of 64 bytes represents the minimum length frame for Ethernet and Fast Ethernet. For 10BASE-T, the slot time is 512 bits × 100 ns/bit or 51.2 μs. For 100BASE-T, the slot time is one-tenth the Ethernet slot time or 5.12 μs. For both Ethernet and Fast Ethernet, a so-called 512 rule defines the network diameter for half-duplex LAN operations. Basically, the 64-byte frame (512 bits) is the smallest packet size that can be transmitted, and the network diameter is determined by the time the shortest frames take to travel round trip to the farthest node in the network. If frames do not arrive by this time, the network can experience late collisions. Thus, the importance of the slot time is twofold. It not only provides time for a station to detect collisions but also specifies the amount of time delay between a transmitting station and a receiving station. Because signals propagate down a transmission medium, this, in turn, results in the slot time governing the network diameter.

DETERMINING ONE-WAY DELAY To determine the one-way delay D_o permissible on an Ethernet LAN, let's consider a two-node network such as the one shown in Figure 5.8. In this example let's assume that station A transmits a frame to station B, requiring a time of T_A to reach

B's adapter. As A's frame traverses the LAN, let's further assume that workstation B has data to transmit. Because A's frame has not yet reached B's workstation, B listens to the LAN, does not "hear" A's frame, and begins its transmission. From a worst-case collision detection perspective, workstation B would begin its transmission just before the first bit in workstation A's frame arrives. This is a worst-case scenario because the collision induces a high voltage that requires the longest time for the frame to flow down the cable and reach workstation A. The high voltage, which indicates the presence of a collision, is applicable only to Ethernet and Fast Ethernet and copper-based Gigabit. However, the Radio Frequency (RF) version of Ethernet, referred to as 10broad-36, uses a different mechanism to detect the presence of a collision. Returning to our two-workstation example, while workstation B will know almost immediately that a collision occurred, this information will take a long time to reach workstation A. If T_A is the time interval for a frame to flow from workstation A to station B and T_B the time for the high voltage to flow from near B to station A, then station A must wait for at least $T_A + T_B$ seconds until it is capable of detecting a worst-case collision.

When the worst-case scenario occurs, we can assume with very near precision that $T_A = T_B = t$. Because the slot time represents the one-way delay, we can express it as a function of total delay as follows:

$$2 \times (\text{slot time}) = t$$

$$\text{or simply as } 2t_s = t$$

This means that the one-way slot time of 512 bits represents half the maximum delay before the network experiences late collisions. Thus, the one-way delay must be doubled to 1024 bit times, which results in an engineering design constraint. While an Ethernet or Fast Ethernet

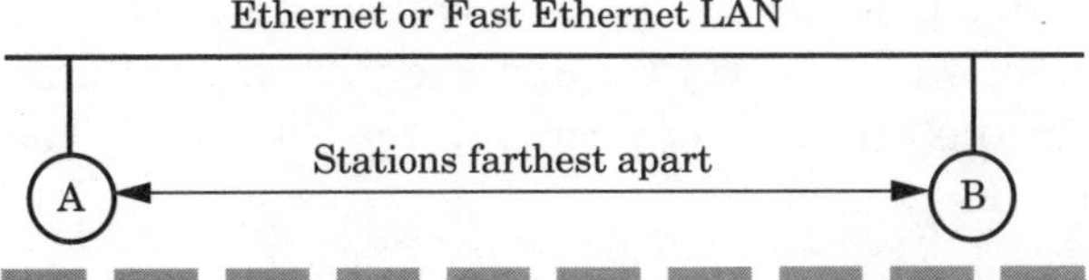

Figure 5.8 The Ethernet and Fast Ethernet network diameter is a function of the propagation delay between the two stations farthest apart on the network.

network will not collapse when this design constraint is exceeded, late collisions will increase. Because collisions result in the execution of a random backoff algorithm, this adversely effects throughput and explains why you should observe network standards.

USING THE VELOCITY OF PROPAGATION We can use the *velocity of propagation* (VoP) to determine the effect of the slot time constraint on the network diameters of Ethernet and Fast Ethernet networks. The velocity of propagation represents the speed of a signal traveling down a cable as a percentage of the speed of light in a vacuum.

The speed of light in a vacuum is 186,000 mi/s (miles per second) or approximately 300,000,000 m/s (meters per second). This is equivalent to 0.3 m/ns or 300 m/μs.

COPPER MEDIA In a copper-medium environment the VoP represents 60 to 70 percent of the speed of light in a vacuum. For category 5 cable, which is used primarily by Ethernet and Fast Ethernet, the VoP is 69 percent of the speed of light in a vacuum. This means that an electrical signal travels along a category 5 cable at 0.3 m/μs × 69 or approximately 0.21 m/ns. For anyone accustomed to thinking in feet, this is equivalent to approximately 8 in/ns (inches per nanosecond). Another method commonly used to express the flow of electrons or photons is to calculate the number of nanoseconds that an electron or photon requires to traverse a meter. This is usually expressed as the inverse of the VoP (V^{-1}), and for electrons in category 5 cable, this is approximately 5 ns/m.

If we apply the preceding metric to a 100BASE-T LAN, we would obtain a value of approximately 127 m for the permissible network diameters, which is 27 percent above the 100 m specified by the standard. The reason for the difference is that the prior computations did not examine other delay factors to include the repetition of a signal entering one port on a hub to the other ports on the hub and the time for an adapter to recognize a collision. Although the preceding computations were not fully comprehensive, they illustrate that the speed of electrons through copper is a key governing factor for an achievable network diameter.

FIBER-OPTIC MEDIA In a fiber-optic environment the speed of photons through the fiber is significantly higher. For example, for multimode fiber photon flow is approximately 3 ns/m. This is 40 percent beyond the flow of electrons in a copper cable and expands the network diameter. Thus, you can use fiber-optic cable to connect stations on a LAN at a considerable distance beyond the 100 m allowed for 10BASE-T and 100BASE-T copper-based-medium LANs. One of the most popular devices that permits an extension of the diameter of a 10BASE-T or 100BASE-T Ethernet LAN is referred to as a *LAN extender*. The actual network diameter achievable depends on several factors. Those factors include the type of optical transmitter and detector used by a pair of extenders as well as the optical cable used to interconnect extenders.

LAN Extenders

There are several types of LAN extenders, each designed to convert an electrical signal to an optical signal for transmission over a specific type of fiber. For example, one popular extender converts 10BASE-T to multimode fiber, extending the transmission distance of the LAN to over 2 km.

Figure 5.9 illustrates the use of a pair of 10BASE-T–multimode fiber extenders. Note that in this example one extender is attached to the port of a hub and uses fiber to enable a remote LAN user to join the LAN even though it is outside the 100-m diameter of the network.

Figure 5.9 Using a pair of 10BASE-T extenders.

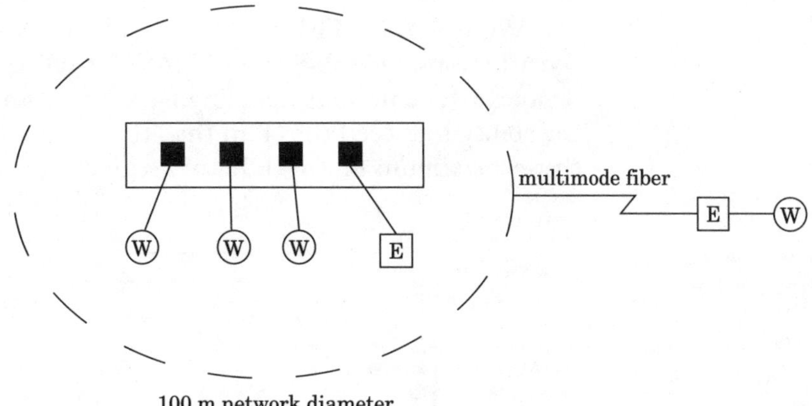

legend
W workstation
E 10BASE-T to multimode fiber extender

FOIRL AND 10BASE-F Two standards govern the transmission of Ethernet across dual-fiber cable: an original standard referred to as the *fiber-optic inter-repeater line* (FOIRL) and the more recent 10BASE-F, which is backward-compatible with the older standard and supports multiple Ethernet connections at distances up to 2000 m (6600 ft) from one another. FOIRL supports distances of up to 1000 m, while 10BASE-F devices double the distance to 2000 m. However, the ability to exceed 1000 m requires the use of 10BASE-F devices at both ends of the fiber.

Under both FOIRL and 10BASE-F standards, an optical transceiver consists of a pulse generating LED, a photodetector, and associated transmit and receive circuitry. Figure 5.10 illustrates a stand-alone repeater, which includes a transceiver, housing, and connectors. Note the two optical connectors that support the connection of two optical fiber cables, enabling one to be used for transmission while the second is used for reception of data.

The physical dimensions of the FOIRL or 10BASE-F repeater/extender are normally equivalent to a pack of cigarettes, and the unit usually costs under $200. A second type of transceiver is commonly built into a fiber hub, permitting several optical ports to be available for use in the hub. In addition to providing optical ports, the fiber hub can contain one or more 10BASE-T ports and an AUI (attachment unit interface) port. Because all ports share a common power supply and circuitry, the cost on a per port basis can be less than the cost of using individual fiber repeater/extenders.

10BASE-FB AND FP In addition to 10BASE-F, there are two related specifications: 10BASE-FB and 10BASE-FP. 10BASE-FB, where the letter "B" denotes a synchronous signaling backbone segment, provides users with the ability to exceed the "4" in the Ethernet 5-4-3 rule. That rule specifies that a maximum of five Ethernet segments can be joined through the

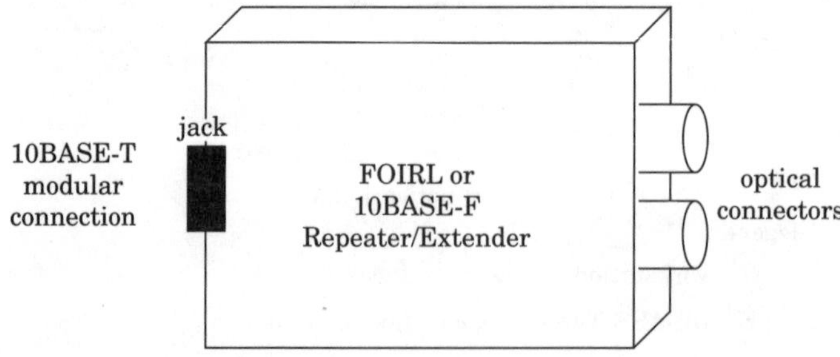

Figure 5.10 A schematic of a FOIRL or 10BASE-F repeater/extender.

use of a maximum of four repeaters, with a maximum of three populated Ethernet segments between two stations on a LAN. A 10BASE-FB signaling repeater is commonly used to connect repeater hubs together in a repeated backbone network infrastructure that can span multiple 2000-m links.

The third version of 10BASE-F is referred to as 10BASE-FP, where "P" denotes the fact that the end segment is a fiber-passive system. Under 10BASE-FP specifications, a single fiber-optic passive-star coupler can be used to connect up to 33 stations that can be located up to 500 m from a hub via the use of a shared-fiber segment.

CABLE SUPPORT Both 10BASE-F and FOIRL support several types of multimode optical cable, ranging from 50/125 through 100/140. Table 5.2 compares optical attenuation values for six types of multimode optical cable. Note that 10BASE-FL has a higher loss budget than FOIRL for each type of fiber, explaining why the transmission distance of 10BASE-FL optical repeaters exceeds the distance obtainable using FOIRL repeaters.

100BASE-FX In the wonderful world of Fast Ethernet, 100BASE-FX is the standard that defines Fast Ethernet transmission over fiber-optic media. The 100BASE-FX standard specifies the use of two-strand 62.5/125-μm multimode optical fiber. The actual encoding of data is based on the use of 4B5B coding, which was first used by FDDI and is also used by 100BASE-TX; the latter requires two pairs of category 5 shielded twisted-pair (STP) cable.

100BASE-FX permits the maximum network diameter of 200 m for 100BASE-TX to be extended to 400 m. However, in a mixed 100BASE-T4–100BASE-FX environment the maximum collision domain should not exceed 231 m, consisting of 100 m for 100BASE-T4 and 131 m for 100BASE-FX. Both ST and SC fiber connectors that were originally defined for FDDI are supported by 100BASE-FX.

TABLE 5.2

10BASE-FL versus FOIRL Optical Attenuation

	Multimode Fiber					
	50/125	**50/125**	**50/125**	**62.5/125**	**83/125**	**100/140**
Numerical aperture	0.20	0.21	0.22	0.275	0.26	0.30
10BASE-FL loss budget, dB	9.7	9.2	9.6	13.5	15.7	19.0
FOIRL loss budget, dB	7.2	6.7	7.1	11.0	13.2	16.5

REPEATER RULES Two types of repeaters are specified for use in a Fast Ethernet environment: class I and class II.

CLASS I REPEATER A *class I* repeater has a greater budget for timing delay, enabling the device to support dissimilar physical media segments that use different signaling methods. The delay for a class I repeater is 140 bit times. In one example of the use of a class I repeater, a 100BASE-T4 segment that uses 8B6T coding is connected to a 100BASE-X segment that uses 4B5B coding. Here the greater timing delay enables a class I repeater to translate the line signal received on one port to the line signal required on the distant segment. Only one class I repeater can be installed in the same segment.

CLASS II REPEATER The second type of Fast Ethernet repeater is a *class II* device. The class II repeater has a lower budget for timing delay; hence it is faster than a class I register. The delay of a class II repeater is 92 bit times. Up to two class I repeaters can be used in a segment, with up to one 5-m interrepeater cable length permitted. Because a class II repeater immediately repeats an incoming signal, it can be used only to interconnect segments that use the same signaling method, such as 100BASE-TX and 100BASE-FX.

SPAN DISTANCE According to the preceding observations, the actual span distance obtainable through the use of repeaters in a Fast Ethernet environment depends on the type of repeaters used and the media employed. Figure 5.11 summarizes Fast Ethernet cabling restrictions based on direct connection and use of different repeaters and media.

Other Extenders

The IEEE 802.3μ Fast Ethernet standard addresses only half-duplex communications. Because it is possible to transmit in a full-duplex mode from a workstation to a switch port, several vendors introduced nonstandardized 100BASE-FX extenders that use both multimode and single-mode fiber in a full-duplex transmission environment. Table 5.3 summarizes the cable distance achievable for 100BASE-TX as well as half- and full-duplex versions of 100BASE-FX, with the latter, as mentioned earlier, not officially contained in the IEEE 802.3μ Fast Ethernet standard.

Figure 5.11 Fast Ethernet distance limitations.

a. Direct connection without using repeaters
(TX = 100 m, TX & FX = NA, FX = 412 m)

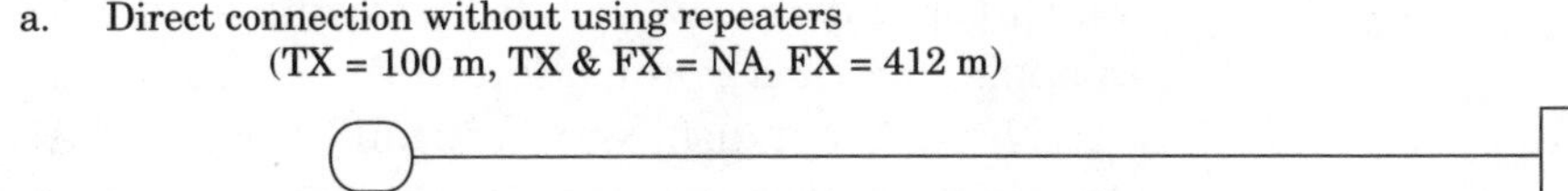

b. One Class I repeater
(TX = 200 m, TX & FX = 260.8 m, FX = 272 m)

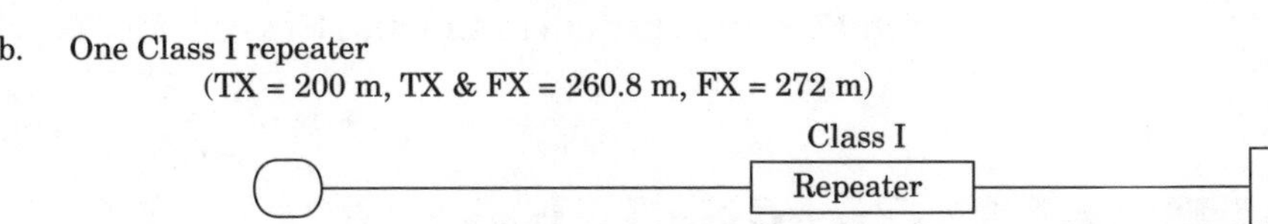

c. One Class II repeater
(TX = 200 m, TX & FX = 308.8 m, FX = 320 m)

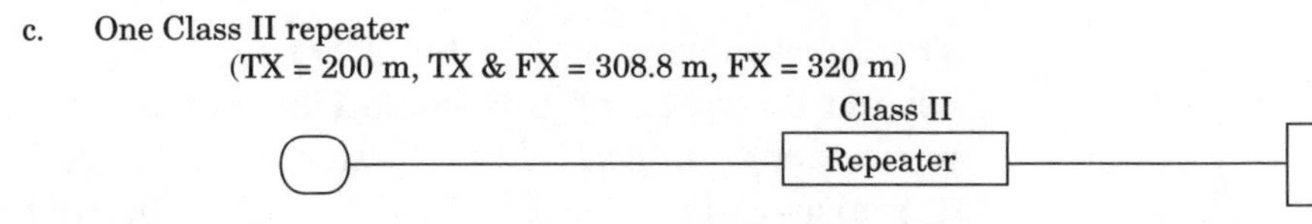

d. Two Class II repeaters
(TX = 205 m, TX & FX = 223 m, FX = 228 m)

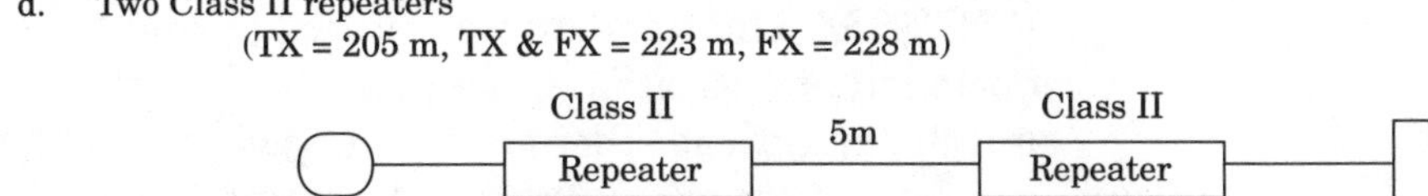

TABLE 5.3 Permissible Fast Ethernet Cabling Distances

Standard	100BASE-TX	100BASE-FX	100BASE-FX	100BASE-FX	100BASE-FX
Media	Category 5 copper	Multimode half-duplex	Single-mode half-duplex	Multimode full-duplex	Single-mode full-duplex
Distance, m 100		412	412	2000	2000

One of the more interesting aspects of the cabling distance achievable with Ethernet and Fast Ethernet extenders is obtained by full-duplex transmission. As indicated in Table 5.3, multimode optical fiber full-duplex transmission permits a cabling distance of up to 300,000 m, while the faster pulse of a laser used with single-mode fiber extends cabling distance to 20,000 m. This makes it possible to connect Ethernet and Fast Ethernet LANs directly to a communication carrier's central office at the LAN operating rate. This also means that it becomes possible to transmit Ethernet and Fast Ethernet at their operating rates, assuming that the applicable fiber is installed to the carrier, enabling the carrier to place the transmission into an applicable SONET or SDH frame for transmission over a WAN. Of course, the distant location must also have an applicable fiber connection to the communications carrier to obtain a LAN-to-LAN transmission capability over a WAN at the LAN operating rate. Now that we generally understand the use of fiber in Ethernet and Fast Ethernet environments, let's consider Gigabit Ethernet.

Gigabit Ethernet

The development of Gigabit Ethernet has a lengthy history that is beyond the scope of this book. However, there are certain milestones worth mentioning as well as the standards that define the technology.

Similar to Ethernet and Fast Ethernet, the IEEE was responsible for standardizing Gigabit Ethernet. The basic Gigabit Ethernet standard is IEEE 802.3Z. This standard was ratified during June 1998 and defines the operation of Gigabit Ethernet over fiber. Another standard for the transmission of Gigabit Ethernet over copper—the IEEE 802.3ab specification, was delayed for several years and became finalized during 2000. Because the focus of this book is on the use of optical components, we will focus primarily on the 802.3z standard.

Overview

Gigabit Ethernet was developed to provide a migration path from 10-Mbit/s Ethernet and 100-Mbit/s Fast Ethernet to gigabit transmission. The primary use of the 1 Gbit/s rate of Gigabit Ethernet is to provide a backbone network infrastructure for organizations running short of bandwidth. Because the per port cost of this technology exceeds that of Fast Ethernet by a factor of >10, it may be quite some time, if ever, before it is used for connecting desktop users. Thus, at the present time Gigabit Ethernet represents a technology oriented primarily toward connecting power stations to a network, such as servers and routers as well as connecting hubs in a tier, two applications that we will examine later in this section.

Gigabit Ethernet was designed to comply as much as possible with previously developed 802.3 Ethernet and 802.3μ Fast Ethernet standards. Thus, Gigabit Ethernet retains the same framing, flow control, and link layer as do earlier versions of this networking technology and makes

Gigabit Ethernet compatible with 10BASE-T and 100BASE-T technologies supported by hubs, switches, and even repeaters. As a result, you can include existing Ethernet and Fast Ethernet network structures with the use of Gigabit Ethernet.

Versions

The IEEE 802.3z specification defines an 8B/10B encoding/decoding scheme that is used by each optical version of Gigabit Ethernet. The 802.3z standard is made up of two optical fiber specifications: one for low-cost copper jumpers or coaxial cable and one for twisted-pair media. Let's first briefly note the meaning of each of the Gigabit Ethernet standards, which define 1-Gbit/s operations over different media. Once this is accomplished, we see how Gigabit Ethernet can be integrated into a network.

FIBER-OPTIC STANDARDS Two fiber-optic standards are associated with Gigabit Ethernet: 1000BASE-SX and 1000base-LX. Because Gigabit Ethernet has a data transmission rate of 1 Gbits/s and a baud or signaling rate of 1.25 GHz, these speeds require the use of laser transceivers instead of LEDs, which have an operational cap of a few hundred megabits per second.

1000BASE-SX The 1000BASE-SX version of Gigabit Ethernet defines the use of an 850 nm laser transceiver operating on multimode fiber. Here the mnemonic SX is used to reference short-wave since the wavelength of the laser is 850 nm. 1000BASE-SX can operate on 50- and 62.5-μm fibers, with a maximum transmission distance of 275 m for 62.5-μm multimode fiber and 550 m when 50-μm multimode fiber is used. The original intention of the IEEE 802.3z 1000BASE-SX standard was to support a transmission distance of up to 300 m for 62.5-μm multimode fiber. Unfortunately, testing of the fiber indicated that the bandwidth of multimode 62.5-μm fiber would not support the target transmission distance under worst-case conditions. Therefore, you may encounter some Web sites as well as trade publications that refer to the *target transmission distance*. In addition to variances in transmission distance based on the diameter of the fiber in micrometers, the transmission distance also varies based upon modal bandwidth expressed in megahertz per kilometer.

In a copper-medium environment bandwidth represents a measure of the information-carrying capacity of the media. In an optical-medium environment other factors affect bandwidth and the transmission capacity of the media. Those factors include the dispersion, or spreading, of light pulses as they travel down the core of a fiber. In a multimode fiber environment, the primary source of dispersion is modal dispersion. Modal dispersion, as indicated earlier in this book, occurs when individual modes of light take different paths through a fiber, thus traversing the length of the fiber at different times. All multimode fibers have a direct correlation between modal spreading and their usable bandwidth; that is, as modal dispersion increases, the usable bandwidth, which is referred to as *modal bandwidth,* decreases. The metric used to define modal bandwidth is MHz·km, which denotes the signaling rate over a specified distance. Thus, any increase in a fiber's modal bandwidth will result in a direct increase in the transmission distance obtainable over a fiber at a given data rate. We will observe this effect when we observe a table of IEEE 802.3z optical media limitations and a figure that plots transmission distance as a function of fiber bandwidth. In addition, after we complete our tour of Gigabit Ethernet specifications, we will tabulate the various flavors of Gigabit Ethernet.

1000BASE-LX A second optical-medium Gigabit Ethernet specification is 1000BASE-LX. Here the mnemonic LX denotes the use of a long-wavelength laser. This specification denotes the use of 1300-nm laser transceivers operating on either single-mode or multimode fiber. When used with 62.5-μm fiber, a transmission distance of up to 550 m can be obtained. Operation over single-mode fiber increases the maximum transmission distance to 3000 m.

COPPER CABLE STANDARDS In addition to supporting the transmission of data over optical fiber, Gigabit Ethernet also supports two specifications for transmission over copper media. Those standards are 1000BASE-CX and 1000BASE-T.

1000BASE-CX The 1000BASE-CX standard is one of two Gigabit Ethernet specifications for transmission over copper cable. The 1000BASE-CX specification covers the use of low-cost copper jumpers operating on high-quality shielded twisted-pair (STP) or coaxial cable. Thus, in all probability CX is a mnemonic for coaxial cable. A transmission distance of up to 25 m is supported by 100BASE-CX.

1000BASE-T The 1000BASE-T specification defines the transmission of Gigabit Ethernet over copper cabling. The actual standard for Gigabit Ethernet is IEEE 802.3ab, which permits a transmission distance of up to 100 m. Table 5.4 summarizes the IEEE 802.3z standard for Gigabit Ethernet over optical media.

ADVANTAGE OF 50-μm FIBER In examining the entries in Table 5.4, we can note the primary advantage associated with the use of 50-μm optical fiber, namely, its higher bandwidth in the shortwave 850-nm operating window. If you look at the modal bandwidth column in Table 5.4, you will note that 50-μm fiber has a modal bandwidth approximately 3 times that of 62.5-μm fiber. Thus, 50-μm fiber permits longer transmission distances at higher operating rates than does 62.5-μm fiber when an 850-nm laser is used.

Actually, 50-μm optical fiber represents a relatively old technology, as it was developed during 1976. While the use of 50-μm optical fiber was common in Germany and Japan, it was relatively slow to materialize in North America. This occurred partly because IBM adopted the newly developed 62.5-μm optical fiber in its cable system in 1986. IBM selected 62.5-μm fiber over 50-μm fiber as it was considered at that time to be better suited for LED light sources. IBM's use was followed by AT&T,

TABLE 5.4

IEEE 802.3z Optical-Medium Limitations

Standard	Fiber type	Diameter, μm	Modal bandwidth, MHz/km	Maximum distance, m
1000base-SX	Multimode	62.5	160	220*
1000base-SX	Multimode	62.5	200	275†
1000base-SX	Multimode	50	400	500
1000base-SX	Multimode	50	500	550‡
1000base-LX	Multimode	62.5	500	550
1000base-LX	Multimode	50	400	550
1000base-LX	Multimode	50	500	550
1000base-LX	Single-mode	9	N/A	3000

*The EIA/TIA 568 building wiring standard specifies the use of 160/500-MHz/km multimode fiber.
†The ISO/IEC 11801 building wiring standard specifies 200/500-MHz/km 50-μm multimode fiber.
‡The specification for the ANSI Fibre Channel is 500/500 MHz × 50-μm multimode fiber.

and 62.5-μm cable was then selected for the FDDI standard; as a result, use of 62.5-μm fiber at 160/500 MHz·km became widespread in North America. However, like Mark Twain, whose rumored demise was greatly exaggerated, 50-μm optical fiber still has life and in fact its use is gaining momentum in North America.

While 50-μm fiber has a significant advantage over 62.5-μm fiber with respect to modal bandwidth and transmission distance, you might also wish to consider its handling and cost. When you do so, as we will shortly note, 50-μm fiber maintains its advantage. Concerning handling, although 50-μm optical fiber has a smaller core diameter than does 62.5-μm fiber, both have similar mechanical strength properties. In fact, both optical fibers have the same glass cladding diameter of 125 μm. Because both types of optical fiber have the same outer diameter, they also have the same general physical characteristics and are similarly priced. Thus, when possible, 50-μm cable represents a better selection than does 62.5-μm cable. In fact, both 50- and 62.5-μm cable are interchangeable. Although there is a one-time attenuation loss that occurs when you couple 62.5-μm fiber into 50-μm fiber, that loss is compensated for by the greater modal bandwidth of 62.5-μm cable. Thus, it is also possible to gradually install 50-μm cable in all premises applications as new requirements occur beyond your existing organizational infrastructure.

VARIATIONS Similar to proprietary versions of Ethernet and Fast Ethernet, there are nonstandardized versions of Gigabit Ethernet. Those nonstandardized versions use 850- and 1300-nm lasers; however, the use of high-bandwidth fiber enables support for transmission distances beyond those noted in Table 5.4. Through the use of fiber that has a bandwidth of up to 1000 MHz/km, it becomes possible to transmit at distances as great as 1300 m under 1000BASE-LX and 625 m under 1000BASE-SX. Figure 5.12 provides a more realistic indication of transmission distance versus fiber bandwidth when 62.5/125 multimode fiber is used.

Duplex Support

Gigabit Ethernet was developed for switched as well as shared media operational environments. Full-duplex operation is considered to represent the preferred mode of operation as it eliminates the possibility of collisions and thus is relatively simple to implement. Full-duplex Gigabit

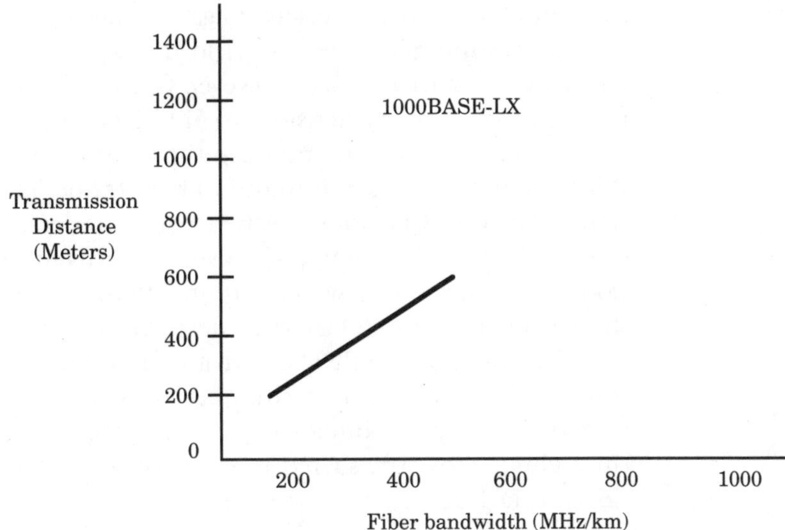

Figure 5.12
Gigabit Ethernet transmission distance.

Ethernet permits simultaneous transmission and reception, permitting a theoretical total of 2 Gbits/s of bandwidth on the same connection. Full-duplex is restricted to a point-to-point networking environment, usually linking hubs together or a hub port to a server. The term "theoretical" is used as a prefix to full-duplex Gigabit Ethernet providing a total of 2 Gbits/s of bandwidth on a connection because of the manner of operation of different devices. For example, if you connect a server to a hub port, the ability to transmit and receive at the same time depends on the activity of workstations accessing the server and the server's response to the queries. Thus, although a maximum data transfer of 2 Gbits/s is possible, it is highly probable that the actual data transfer rate will be considerably below the theoretical data transfer rate.

In a half-duplex operating environment, Gigabit Ethernet uses the CSMA/CD protocol, similar to Ethernet and Fast Ethernet. However, to maintain a 200-m collision diameter for half-duplex operations, the minimum CSMA/CD carrier time and Ethernet slot time were extended from 64 to 512 bytes. The extension occurs through the use of carrier extension symbols as shown in Figure 5.13. In examining this figure, note that the minimum Gigabit Ethernet frame length is still 64 bytes, which maintains interoperability with Ethernet and Fast Ethernet.

Because the Ethernet slot time is extended from 64 to 512 bytes, half-duplex Gigabit is ill-suited for transmitting queries, which typically are short in nature. This is because Gigabit Ethernet will first insert pads to form a minimum 46-byte data field in the same manner that Ethernet and Fast Ethernet operate. However, Gigabit Ethernet will then extend the frame via carrier extensions to 512 bytes. Thus, the transmission of a 10-byte query would first result in the addition of 36 pads to a relatively short frame to form a minimum 64-byte frame. Then, 448 carrier extension bytes would be added. When you consider the fact that each version of Ethernet has 26 bytes of overhead, this means that the transmission of a 10-byte query would require 510 bytes. This also explains why a shared media Gigabit Ethernet network may provide a transmission capacity only marginally above that of Fast Ethernet when most of the transmission carries interactive queries. Now that we are beginning to understand why full-duplex Gigabit Ethernet can provide a significant throughput above the shared media version of this technology, let's consider the technology in a network.

Network Utilization

The primary use of Gigabit Ethernet is in a switched LAN environment, where the switch contains either a mixture of Gigabit Ethernet ports and other types of Ethernet and/or Fast Ethernet ports or Gigabit Ethernet ports only. Figure 5.14 illustrates the use of Gigabit Ethernet in a tiered network switch environment. In this networking environment several Ethernet/Fast Ethernet switches are used to provide communications capability to different departments within an organization. Note that each departmental switch uses a Fast Ethernet connection from the switch to a departmental server. Also note that one port on each departmental server uses a gigabit connection to an enterprise switch. That switch, which is shown at the top of the tiered switching hierarchy in

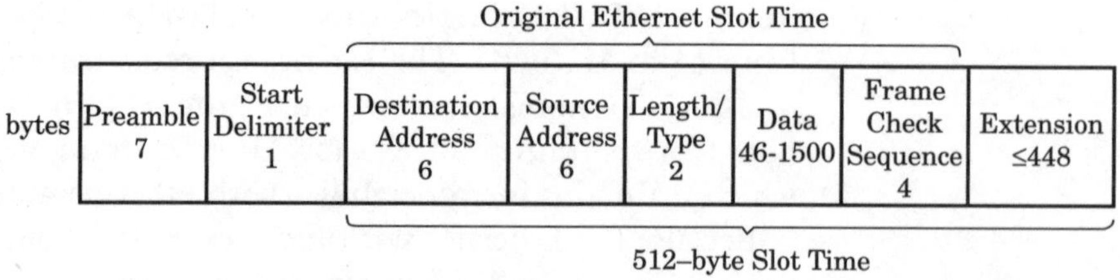

Original Ethernet Slot Time

	Preamble	Start Delimiter	Destination Address	Source Address	Length/Type	Data	Frame Check Sequence	Extension
bytes	7	1	6	6	2	46-1500	4	≤448

512–byte Slot Time

Figure 5.13 Half-duplex Gigabit Ethernet uses a carrier extension scheme to extend timing so that the slot time consists of at least 512 bytes.

Figure 5.14, represents a true Gigabit Ethernet switch with all ports on the switch operating at 1 Gbit/s. Note that enterprise servers are connected to this switch, which enables employees at different locations within the organization to rapidly access enterprise servers. Also note that because numerous servers are clustered together at one location, this networking strategy results in a group of servers connected to a common switch, which is referred to as "establishing a server farm."

Limitations

No discussion of Gigabit Ethernet would be complete without mentioning two of its key limitations. Those limitations are the bus speed and the disk transfer capacity of many servers, which cannot keep pace with the transmission rate of the technology and the lack of redundancy.

BUS SPEED AND DISK TRANSFER To appreciate the importance of bus speed, consider Table 5.5, which lists the throughput in megabytes

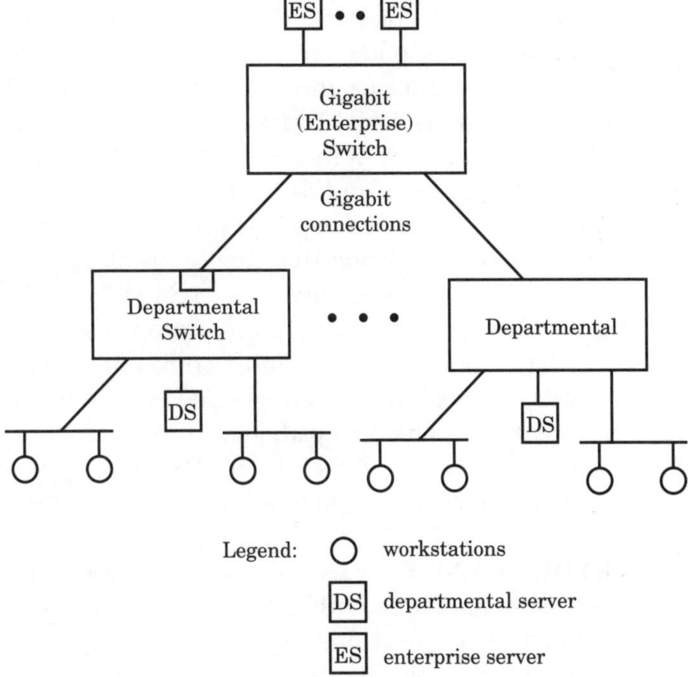

Figure 5.14 Using a Gigabit switch in a tiered network environment.

TABLE 5.5

Theoretical Bus Throughputs

Bus	Width, bits	Throughput, Mbytes/s
ISA	16	8
EISA	32	33
MCA	32	20
PCI, 33 MHz	32	132
PCI, 33 MHz	64	266
PCI, 66 MHz	32	266
PCI, 66 MHz	64	533

per second for seven types of PC buses. In examining the entries in Table 5.5, it is obvious that with the exception of the Peripheral Component Interconnect (PCI) bus, Industry Standard Architecture (ISA), Extended Industry Standard Architecture (EISA), and Microchannel Architecture (MCA) buses do not have the capability to transfer data at a rate anywhere near 1 Gbits/s. Even when a 32-bit PCI bus is used, its theoretical throughput is 132 Mbytes × (8 bits/byte) or 1.056 Gbits/s. However, because a computer needs cycles to retrieve data from a disk and initiate an adapter transfer, most PCI-based PCs can achieve a sustained data transfer only under 500 Mbits/s. However, modifications to Windows NT and the use of Windows 2000 with high-performance 1 GHz Pentium III processors and 64-bit PCI adapters operating at 66 MHz have resulted in a sustained data transfer of 900 Mbps.

When contemplating the installation of a Gigabit Ethernet switch-to-server connection, it is important to carefully consider both the data path and the clock rate of the PCI bus. Unfortunately, the 32-bit data path and 33-MHz clock rate is the most common bus found in workstations and servers. Fortunately, and for added cost, some servers can be obtained with either a 64-bit data path and 33 MHz of clock speed or a 64-bit data path and 66 MHz of clock speed. The latter type of system provides the capability to effectively support Gigabit Ethernet transmission.

REDUNDANCY A second major limitation of Gigabit Ethernet is its lack of redundancy. Unlike FDDI, which provides built-in redundancy through a self-healing ring, Gigabit Ethernet does not, and a break between a Gigabit Ethernet switch and another switch or file server can be very disruptive to a company. Recognizing this problem, some companies introduced dual-port network adapters with an automatic cutover capability. While such adapters are more expensive than single-port adapters, for organizations that require redundancy, the added cost may be well worth reaching deeper into one's pocket.

CHAPTER 6

Fiber in the WAN

In this chapter we discuss the use of fiber in the wide area network (WAN). First, we will briefly discuss the evolution of the use of fiber as a mechanism to provide high-speed communications between communication carrier central offices. We will also discuss the rationale behind the use of fiber in the WAN. Once this is accomplished, we will turn our attention to the development of standardized optical transmission systems in the form of SONET in the United States and the Synchronous Digital Hierarchy (SDH), the European equivalent of SONET. After examining the characteristics and operation of SONET and SDH, we will then focus on two techniques that were developed to significantly increase the transmission capacity of a fiber: wavelength division multiplexing (WDM) and dense wavelength division multiplexing (DWDM).

Evolution and Rationale

The use of optical fiber in the WAN dates to the 1980s, when Sprint and AT&T began installing fiber throughout their backbone networks. In 1983 AT&T installed its FT3C lightwave transmission system in what was then referred to as the Northeast Corridor Project, with fiber routed from Boston to Washington, DC.

The AT&T FT3C lightwave transmission system is notable because it used an early version of wavelength division multiplexing (WDM) to transport three 90-Mbit/s signals on a common fiber. This transmission method enabled over 240,000 simultaneous voice conversations to be carried on one cable, which was a record for its time. In fact, the AT&T WDM system can be considered to represent a very early version of WDM. That system was based on the use of LEDs and multimode fiber, with the output of three LEDs operating at different frequencies coupled onto a common fiber. The result was the transmission of three data streams, each operating at 90 Mbits/s on distinct optical frequencies. Thus, the AT&T FT3C lightwave transmission system seems to be based on the concept of frequency division multiplexing. By early 1991 Sprint became the first of the three major communications carriers to completely convert their intercity communications from microwave to fiber, quickly followed by other long-distance carriers.

Although the initial long-haul fiber systems were based on the use of multimode fiber and LEDs, the demand for higher operating rates resulted in a trend toward the use of single-mode fiber and lasers. As previously noted in this book, LEDs have a maximum data transmission capability near 300 Mbits/s, which is significantly below the capability of lasers.

The use of single-mode fiber and lasers resulted in a significant increase in the transmission capacity of optical systems. Today data transmission rates of 10 Gbits/s, are commonly employed in the backbone infrastructure of many long-distance communications carriers, and 40-Gbit/s systems are being used in field trials and may be in commercial service by the time you read this book.

Rationale for Use

The growth in the use of fiber for intercity communications can be summed up in three words: "capacity" and "error rate."

CAPACITY Concerning capacity, as noted earlier, the AT&T FT3C lightwave system supported 240,000 simultaneous voice conversations. To put this capacity in perspective, let's convert it to its equivalent number of T1 and T3 circuits.

A T1 circuit is used to transport 24 voice conversations. Thus, the AT&T FT3C fiber system has the capacity equivalent of 240,000/24 or 10,000 T1 circuits! Yes, that's 10,000 T1 circuits! Concerning T3 circuits, first note that a T3 circuit transports 28 T1s. Since a T1 transports 24 voice conversations, this means that a T3 transports 28 × 24 or 4672 voice conversations. Thus, the AT&T FT3C lightwave system has the capacity of 240,000/672 or 357 T3 circuits!

In the previous computations we can note that the AT&T FT3C fiber system represented by a thin piece of glass surrounded by shielding replaces 10,000 copper pair wires used to provide a T1 transmission capability or 357 coaxial cables which are used when T3 circuits are installed on copper conductors. If you think about the physical wiring required for those T1 or T3 circuits, their power requirements, and the need to install repeaters with power and to shelter approximately every 6000 ft, you can begin to appreciate the cost savings associated with the use of optical fiber.

Although the preceding computations are accurate, they do not represent the true state of the present capacity of fiber in comparison to T1 and T3 copper circuits. This is because there have been significant advances in the design of lasers that pulse at higher rates, permitting an increase in transmission rates at a given frequency. As we noted earlier in this chapter, optical systems based on the use of single-mode fiber and lasers commonly operate at 10 Gbits/s, and field trials of systems operating at 40 Gbits/s are now being conducted. When coupled with advances in wavelength division multiplexing (WDM) and dense wavelength division multiplexing (DWDM), the capacity of an optical fiber has increased by several magnitudes and explains the popular Qwest Communications commercial in which a person entering a store is told that that company can transmit the contents of every book written in every language in a small amount of time.

BIT ERROR RATE Earlier in this book we noted that one of the key advantages obtained from the use of optical fiber is the immunity of light transmission to electromagnetic radiation. This not only allows fiber-optic cable to be routed through locations close to machinery but also significantly lowers the error rate resulting from transmission occurring on optical fiber in comparison to the use of copper cable.

Many sources of impairments (interference) adversely affect copper cable, ranging in scope from crosstalk to impulse noise caused by lightning and machinery. Optical fiber is immune to those impairments, which commonly results in an error rate several orders of magnitude lower than on copper cable. Perhaps the only common impairment that affects both copper cable and optical fiber is Joe Backhoe, who typically works where he shouldn't. When Joe cuts a cable, then you unfortunately will note another difference between copper cable and optical fiber. That difference is the number of adversely affected users, which is higher when an optical fiber is cut since the fiber supports many more users than does a copper cable.

The AT&T FT3C System

The AT&T FT3C lightwave transmission system can be considered as a milestone in the deployment of optical fiber. In addition to providing a showcase high-speed transmission system between Boston and

Washington, DC, the FT3C system represented the first commercial use of wavelength division multiplexing. To understand how this was accomplished, consider Figure 6.1. *Frequency division multiplexing* (FDM) is illustrated in the top portion of the figure. Note that FDM represents an analog technology in which the bandwidth of the transmission medium is subdivided by frequency. In the top portion of the figure three distinct frequencies (f_1, f_2, and f_3) are used to derive three separate channels. Each channel can be modulated separately from the other channels, permitting three simultaneous transmissions to occur on one analog circuit. In fact, most trunks linking telephone company central offices used FDM through the 1960s to transport multiple voice conversations over one physical line connecting geographically separated offices.

Perhaps thinking about the manner by which FDM operated permitted engineers insight into developing WDM equipment. Early versions of WDM coupled light from two or more sources at discrete wavelengths into a fiber, with each wavelength in effect becoming a

Figure 6.1 Frequency division multiplexing and wavelength division multiplexing.

a. Frequency Division Multiplexing

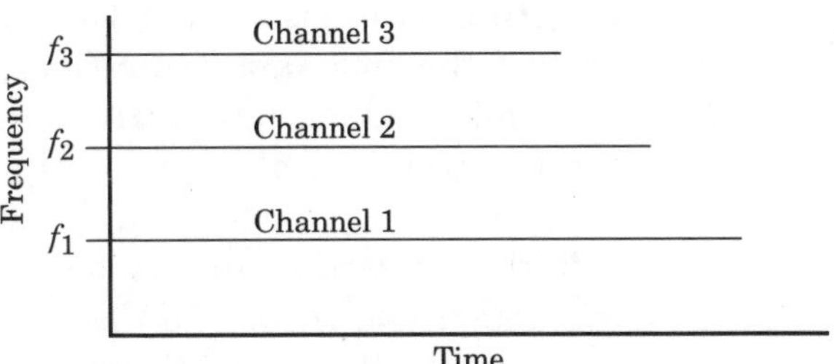

b. Wavelength Division Multiplexing

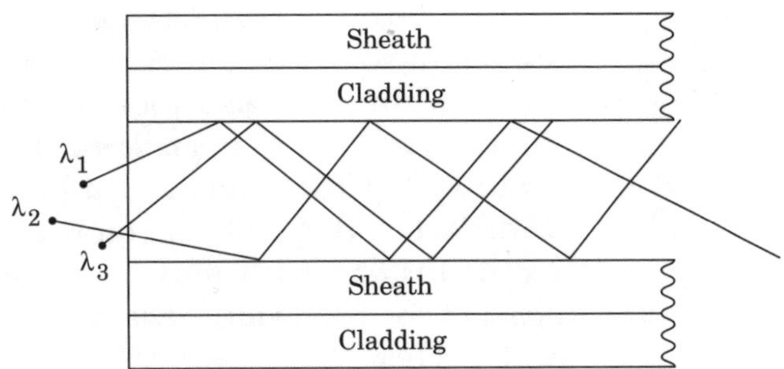

transmission channel operating at the modulation rate used by the equipment driving the light source. The lower portion of Fig. 6.1 illustrates the use of three light sources at distinct wavelengths to form three channels within an optical fiber. As we noted earlier in this chapter, the AT&T FT3C lightwave system coupled the output of three LEDs onto a common multimode optical fiber, representing the first commercial use of WDM.

Advances in Technology

Although the AT&T FT3C lightwave transmission system was a pioneering effort, it was rapidly rendered obsolete by advances in optical technology. For example, while AT&T was constructing its FT3C system, British Telecom demonstrated the use of a single-mode fiber capable of carrying signals at 10 times the rate of the AT&T system. Within a few years Sprint, AT&T, and MCI were using single-mode fiber in their backbone networks as it provided a transmission capacity much greater than the AT&T pioneering system.

As the three major communications carriers wired their infrastructure with "glass" fiber, fiber was extended to link the continents. In 1988 the first submarine fiber cables were laid across the Atlantic, to be followed the next year by fiber cables across the Pacific. Within a few years optical transmission rates reached 2.5 Gbits/s, which appeared to represent a sufficient transmission rate that could provide the necessary capacity to interconnect cities and continents for the foreseeable future. Similar to the play titled *A Funny Thing Happened on the Way to the Forum,* a pair of events transpired that made the projections of pundits obsolete. First, the digital communications revolution occurred during the mid-1980s, with tens of thousands of bulletin board systems serving millions of PC users. Instead of a typical 3-min voice call, a person accessing a bulletin board system commonly maintained access for 20 or more minutes, requiring communications carriers to upgrade their facilities. Just when those carriers probably felt they were ahead of the curve, along came the World Wide Web and the Mosaic browser during the mid-1990s. The rest is recent history, with tens of millions of Internet users driving a demand for bandwidth that could be satisfied only from the use of optical fiber.

INITIAL WAN ARCHITECTURE The initial architecture associated with the use of optical fiber in the WAN remained a sequence of point-to-point transmission systems until the late 1990s. Electrical signals would be converted by a laser at one end of a fiber, while a photodetector at the other end would convert the light signals back into an electrical form. Once converted back into an electrical format, the electronics could process and switch signals among different paths or simply serve as a temporary conversion point for regeneration of the signal. Thus, the optical-to-electrical conversion followed by an electrical-to-optical reconversion functioned as an amplifier.

BROADBAND WDM The first development of WDM based on the use of lasers is referred to as *broadband WDM.* Broadband WDM took advantage of the fact that it was relatively easy to use lasers that operate in the 1300- and 1500-nm optical windows and couple their output onto a common single-mode optical fiber. During 1994 the output of lasers operating at 1310 and 1550 nm were combined onto a common optical fiber. Each laser operated at 2.5 Gbits/s, resulting in a transmission capacity of 5 Gbits/s obtained on a single fiber. Although a transmission capacity of 5 Gbits/s does not begin to compare to the transmission capacity obtainable just a half decade later, at the time it provided communications carriers with an inexpensive alternative to installing new fiber to meet customers' expanding transmission requirements. Thus, instead of having to physically install a new optical fiber, it became possible to add a second laser input and a coupler to double system capacity. Similar to a child in a candy store, communications carriers were able to taste the economic advantages associated with broadband WDM and provided a ready market for more modern systems.

WDM AND DWDM Broadband WDM can be considered as laying the foundation for advances in the development of techniques to enhance the transmission capacity of optical fiber. As optical filters and laser technology improved, it became both possible and practical to combine additional signals on a fiber. In 1995 the number of channels, each consisting of a separate optical signal at a distinct wavelength, increased to 4 per fiber. Since then advances in technology resulted in the number of channels per fiber increasing to 8, then 16, 32, 40, 64, and even 80, with systems with 128, 256, and even more channels being developed, as we note later in this chapter.

As the number of channels in an optical multiplexing system increased, the term *dense wavelength division multiplexing* (DWDM) came into use in reference to such systems. While there is no magic number of channels where a WDM system becomes a DWDM system, today most optical multiplexing systems are classified as DWDM systems.

SYSTEM CAPACITY We can obtain an indication of the growth in the transmission capacity of optical fiber systems by focusing on the technology at certain periods of time. To do this, let's commence our discussion with the AT&T FT3C lightwave transmission system that became operational in 1983. As noted earlier, the FT3C system provided a composite transmission rate of 270 Mbits/s.

In 1994, as previously noted, a broadband WDM system was developed through the use of two 2.5-Gbit/s lasers. This system provided a composite transmission capacity of 5 Gbits/s. In 1995 an increase in the number of channels to 4, each operating at 2.5 Gbits/s, resulted in a composite transmission rate of 10 Gbits/s.

If we fast-forward our timeline to the beginning of the new millennium, by the year 2000 a transmission rate of 10 Gbits/s was obtainable on a per channel basis. With 32 and 40 channel systems representing common offerings, the composite transmission rate obtainable on a single optical fiber increased from 320 to 400 Gbits/s. During the year 2000, field trials using lasers operating at 40 Gbits/s were in progress. At the same time work was proceeding to extend transmission rates to 40 Gbits/s, work was also progressing to add more wavelengths to a DWDM system. To illustrate the potential growth in the transmission capacity of a single optical fiber, let's examine the use of evolving 128- and 256-channel DWDM systems. At 10 Gbits/s per channel, a 128-channel system provides a composite transmission capacity of 1.28 Tbits/s! Yes, that's 1.28 trillion bits per second, a transmission capacity sufficient to carry all data, voice, and video generated by every person on the planet as recently as approximately 1991. If we sharpen our pens and perform a simple computation, we note that the transmission capacity of a single optical fiber has increased by a factor of 1.28 Tbits/s to 5.12 Tbits/s, or approximately 474,000 times from AT&T's FT3C lightwave system introduced during 1983. Note that this represents an increase in transmission capacity of over six magnitudes!

If we consider the possible near-term availability of 256-channel systems, with each channel operating at 40 Gbits/s, the capacity of a single optical fiber may appear to resemble a science fiction fantasy. For example, with 256 channels, each operating at 40 Gbits/s, the composite capacity of a single optical fiber becomes 10.24 Tbits/s.

Instead of being science fiction, the increase in the transmission capacity of optical fiber is a reality. However, this increase resulted from three technological advances. Those advances include the manufacture of better-quality glass fibers, the development of fiber Bragg gratings, and the availability of optical amplifiers, three topics that we now focus on.

ADVANCES IN OPTICAL FIBER The transmission of multiple wavelengths with high levels of optical power can result in several nonlinear phenomena that can adversely affect the performance of a transmission system. Some of the factors that can adversely impact optical transmission include impurities in the optical fiber that result in dispersion of the components of a signal as well as such nonlinear effects as four-wave mixing, modulation instability, and self-phase and cross-phase modulation.

Signal distortion resulting from signal dispersion can be limited by selecting purer fiber as well as by using a compensating device with a polarity opposite to that of the transmission fiber. Although such techniques reduce dispersion, they do not reduce the effect of nonlinear phenomena that can also adversely affect the performance of an optical transmission system. Concerning nonlinear effects, four-wave mixing (FWM) occurs when multiple wavelengths are transmitted over a common fiber. Modulation instability results from the interaction between an optical signal and an optical amplifier. When multiple wavelengths are transmitted on a common fiber, modulation instability results in individual wavelengths mixing, generating new, unwanted wavelengths. Those unwanted wavelengths can both interfere with the original signals as well as mask a portion of their power, making it harder on the receiver to discriminate the correct signal from the unwanted wavelengths. Two additional nonlinear effects are self-phase and cross-phase modulation. *Self-phase modulation* represents a term used to describe the broadening of light pulses as they flow down a fiber. In a WDM or DWDM system, a similar effect is referred to as *cross-phase modulation.*

In an effort to counter the effects of dispersion and nonlinear phenomena, several optical fiber manufacturers developed new products. One such product was a so-called *non-zero-dispersion-shifted large-effective-*

area fiber (LEAF) developed by Corning. This so-called LEAF optical fiber has a low dispersion coupled to a large effective area A_{eff}. The term *effective area* represents the average area of a fiber in which optical power is transmitted. For example, when using single-mode fiber, the effective area is approximately equivalent to the fiber core area. The larger the effective area, the greater the amount of optical power that a fiber becomes capable of supporting without becoming susceptible to the nonlinear effects mentioned above. Thus, LEAF fiber's large A_{eff} provides higher power handling capability as well as a degree of immunity to those nonlinear effects. For this reason, LEAF and similar fiber from other optical fiber manufacturers permits extended transmission distances before light signal amplification is required. Because LEAF reduces the above-mentioned nonlinear effects, it also provides a medium more conducive to the development of additional channels on a DWDM system. Now that we have an appreciation for the role of improved fiber in the development of optical systems with greater transmission capacity, let's study a second area that makes DWDM a reality. That area is fiber Bragg gratings.

FIBER Bragg GRATINGS *Fiber Bragg gratings* represent an important component that enables modern optical multiplexing to occur. A fiber Bragg grating represents a small section of a fiber that is modified to create periodic changes in the index of refraction. Because it is possible to vary the space between the changes in the index of refraction, it becomes possible to create an optical filter. Thus, by varying the space between the different index of refraction created in a fiber, a certain frequency of light, which is referred to as the *Bragg resonance wavelength,* is reflected back while all other wavelengths pass through the fiber. With fiber Bragg gratings it becomes possible to develop optical add/drop multiplexers. This, in turn, allows specific wavelengths to be added or removed from an optical fiber. To understand why this capability represents another technological development that enhances the growth in DWDM systems, assume that a communications carrier intends to install a multiplexed optical transmission system between New York and Miami. Through the use of fiber Bragg gratings, it becomes possible to drop a channel in Baltimore, another channel in Washington, DC, and a third channel in Richmond, Virginia. In addition, if traffic warranted it, channels could also be added to the fiber at those or other locations along the route from New York to Miami. Now that we have an appreciation for the role of fiber Bragg gratings, let's discuss a third area of technology that enables WDM and DWDM to become a reality. That area of technology is erbium amplification.

ERBIUM AMPLIFICATION Optical transmission is similar to electrical transmission in that both types of signaling become weaker and distorted as transmission distance increases. In an optical environment, which in many ways is similar to a copper cable environment, the ability to extend transmission requires signals to be periodically amplified. Light pulses can typically travel between 40 and 80 km before becoming too attenuated and dispersed.

Until the late 1980s the only method available to amplify a light signal was through the use of an electronic repeater. In this method, the repeater performed an optical-to-electrical and electrical-to-optical conversion as illustrated in Figure 6.2. In effect, this action regenerated a new optical signal, and the device was referred to as an *optical regenerator.*

The use of an optical regenerator considerably adds to the cost of a long-distance fiber. Thus, the realization that erbium has an electron shell or band whose energy-level difference from the base configuration of erbium ions, referred to as a *metastable band,* is close to the energy of a photon in the 1550-nm range, permitted this rare-earth element to serve as an optical amplifier. The rationale for the use of erbium is the fact that the 1550-nm region represents a midpoint position in an optical "window," where signal loss values are lowest in that optical region. The optical region between 1530 and 1565 nm is referred

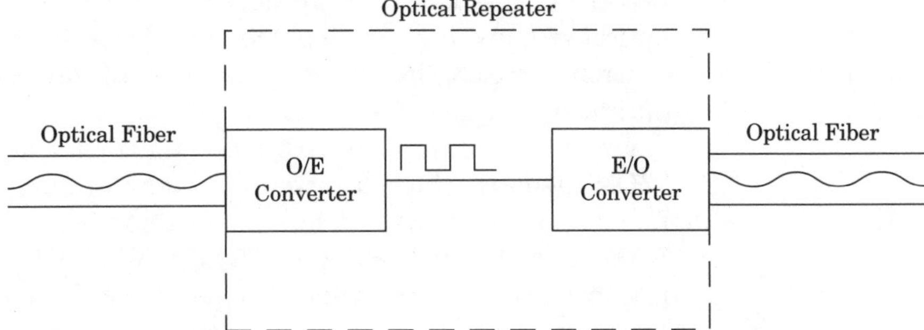

Figure 6.2
The optical repeater represents an optical-to-electrical and electrical-to-optical converter that regenerates an optical signal.

to as the *C band,* and represents the third well-known telecommunications window.

An erbium ion exposed to an intense level of light at certain wavelengths will absorb photons, while some of its electrons will jump to a higher-energy metastable orbit. After an electron jumps to the higher-energy band, it returns to the base band and gives up a photon that has the same wavelength and phase as the incoming photon. Thus, the doping of erbium acts as an optical amplifier since the incoming photons result in a cascade of new photons with the same direction, wavelength, and phase.

In addition to the doping of a length of fiber with erbium, a power source is required to provide the energy for erbium electrons to jump to a higher-energy metastable orbit. That power source is a pump laser that commonly operates at 980 nm. Thus, the pump laser provides the energy for the amplification process. With the use of *erbium-doped fiber amplifiers* (EDFAs), it becomes possible to have optical fiber routed for distances from hundreds to thousands of kilometers at operating rates as high as 10 Gbits/s without requiring an electronic optical repeater. Because the EDFA replaces the electronic optical repeater, significant savings can be obtained when a long-haul fiber is installed. Although a "hut" is still required approximately every 80 km to contain an EDFA and pump laser, the size, complexity, and cost of equipment in each hub is less than when electrical optical repeaters were previously required.

As research progressed following the development of EDFA, this method of optical amplification was determined to have the potential to simultaneously boost the power of many closely spaced optical channels. Thus, EDFA technology opened the door to the ability to design optical systems that could contain an increasing number of channels at different wavelengths, with the power of all the signals boosted by one optical amplifier.

One practical result of EDFA was the ability to develop economical WDM systems, which, as a result of erbium doping, had an extended range prior to requiring conventional amplification. After the year 2000, WDM quickly became dense -WDM (DWDM). In fact, the ITU (International Telecommunication Union)-standardized DWDM by defining a grid of wavelengths that we discuss later in this chapter. Because of the demand for an increased level of transmission capacity, vendors introduced proprietary systems that use as few as 0.8 nm sections of frequency, enabling 80 channels to be created in the frequency spectrum associated with the main erbium-doped fiber amplifier (EDFA) band. Shortly thereafter a second EDFA band was used, making it possible to derive an additional 80 channels in the 1570- to 1610-nm band, thus doubling the capacity to 160 channels.

This band is referred to as the *L band,* and the use of erbium-doped amplifiers in this band is referred to as *L-band EDFA.* Although the idea behind L-band EDFA dates to the early 1990s, the fraction of erbium ions that can be excited in the C band, referred to as *ion inversion,* is between 70 and 80 percent, while its use in the L band provides an inversion of approximately 40 percent. This results in L-band amplification approximately 20 percent that of a C-band EDFA, requiring more amplifiers to be inserted at an additional cost when the L band is used.

Not quite content with 160 channels, Lucent Technologies as well as other vendors, according to trade press reports, were on the verge of introducing systems that could support over 1000 distinct wavelengths for transmission over a single fiber when this book was written. However, it was not stated what band would be used or the gain that could be expected through the use of EDFAs.

While EDFA was found suitable for amplification of signals up to 10 Gbits/s, as vendors experimented in extending data transmission rates to 40 Gbits/s, it became clear that system noise was a major limitation associated with developing very-high-speed optical networks. At 40 Gbits/s a maximum transmission distance of approximately 40 km was obtainable prior to requiring signal regeneration. Because the existing long-haul infrastructure of most communications carriers includes huts where power supplies and electronics were previously installed to support optical amplification, from an economic perspective it is important to be able to use the existing infrastructure. To accomplish this requires a new method of amplification at a rate of 40 Gbits/s. That method of amplification is referred to as *Raman amplification.*

RAMAN AMPLIFICATION A *Raman amplifier* is based on a phenomenon referred to as the *Raman effect in optical fibers.* The Raman effect occurs when a large continuous-wave laser signal is colaunched at a lower wavelength than the signal to be amplified, resulting in a gain whose value is dependent primarily on the strength of the laser pump and the frequency offset between the pump and the signal. When the pump photon gives up its energy to create a new photon, some residual energy is also created. The residual energy is absorbed in an optical fiber as vibrational energy that results in gain or amplification of the signal to be amplified. The key to Raman amplification capability is the launching of pump light into a fiber at amplifier sites opposite the signal direction. This action results in Raman amplification functioning as a low-noise preamplifier that maintains signal integrity. In addition, because the signal power is weaker at the output end of a fiber, no additional nonlinear fiber phenomena are introduced.

Another advantage associated with Raman amplification is the fact that an existing optical fiber can be used as a medium when properly pumped. This means that amplification is not dependent on doping of the fiber and permits the use of a wider range of existing fiber.

SIGNAL SWITCHING Another evolving optical area that warrants attention concerns signal switching. Today the switching of an optical signal is primarily an all-or-nothing affair. That is, if Joe Backhoe digs where he shouldn't and breaks a SONET ring, within a few milliseconds the break will be detected and the full optical flow will commence in an opposite direction to provide a restored communications capability. To be truly effective requires a higher level of optical switching. For example, assume that a DWDM fiber is routed from Boston to Washington, DC through New York and Philadelphia. Instead of using optical-to-electrical converters at each city to feed an electronic switch, a far simpler mechanism and obviously less expensive one would be to drop wavelengths at the intermediate locations. While some equipment permits communications carriers to drop a wavelength, most carriers would like the ability to both drop and add wavelengths at intermediate locations. The key to obtaining this capability is a true optical switch, which would direct signals from any input port to any output port as illustrated in Figure 6.3.

Optical cross-connections currently occur using digital electronics, requiring the incoming optical signals to be converted into electrical pulses. One emerging type of optical switch being developed consists of arrays of miniature mirrors that move back and forth to redirect light. This type of system is referred to as a *microelectromechanical system* (MEMS). A relatively recent MEMS development includes the etching of planar arrays of tiny components to form silicon substrates. Through the etching process mirror elements become suspended on tiny posts

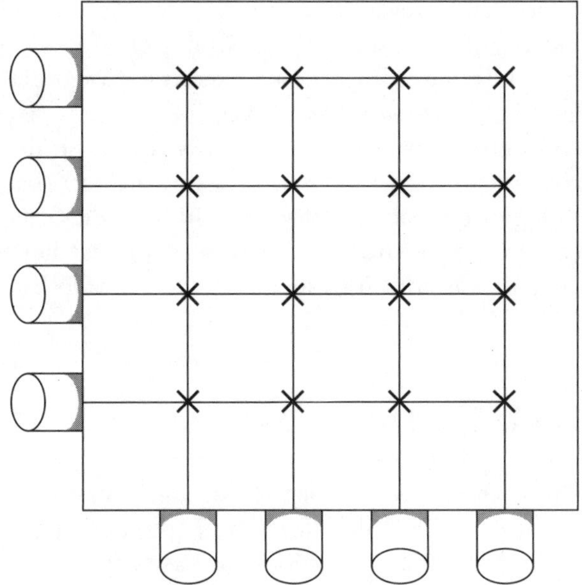

Figure 6.3
A true optical switch will direct wavelengths from within each fiber input on one port onto a fiber on an output port. Attempts to create an optical switch include the use of pop-up and other mirrors, etching of planar arrays, and the use of intersecting waveguides.

above the substrate. This permits a voltage to be applied to circuitry on the substrate, which in effect pulls on the side of a mirror, tilting the mirror at an angle commensurate with the applied voltage to reflect an input beam to an output port. Because this method of switching is electrically driven, it is faster than the MEMS method, which is an electromechanical method.

A third design being tested employs pop-up mirrors. This design is referred to as a *digital MEMS* switch since each mirror can be placed in one of two possible positions, either flat along the surface or in a pop-up position. In the pop-up position the mirror intercepts a light beam and reflects it to an applicable output port. Although this switching method is less likely to misdirect a beam of light, the design mechanism is more complex.

A fourth method of optical switching being considered involves the use of two intersecting sets of parallel waveguides set in a solid substrate. Similar to the composition of an optical fiber core, the waveguide layers have a higher refractive index than does the surrounding material; consequently, the light flows within the waveguide. This method also involves placement of a tiny set of liquid-filled holes at intersecting points. When the holes are filled with liquid, light passes through the intersection. However, draining the hole results in the formation of a bubble at the intersection, causing a change to the refractive index, which reflects light into the intersecting waveguide. This type of switch is referred to as a *microbubble switch.*

Although it may be several years until the deployment of optical switches is commonplace, it is probably a given that one or more of the technologies described above will illuminate the future. Now that we have a background in the direction of optical networking in the WAN, let's turn our attention to specifics, namely, SONET and SDH.

SONET and SDH

Although the use of optical fiber significantly increased during the 1980s, one key difficulty faced by equipment purchasers was a lack of standards. Most optical equipment was proprietary, which precluded the ability of communications carriers to mix equipment from different vendors. In addition to creating interoperability problems, the lack of standards also resulted in the inability of carriers to obtain multiple bids for additional equipment after a primary vendor was selected. This had the effect of locking a communications carrier to a specific optical component vendor.

Recognizing the problems mentioned above, the Exchange Carriers Standards Association (ECSA) initiated action during 1984 to develop a standard to govern the connection of one optical fiber system to another. While this effort governed the interconnection of optical fiber systems, it wasn't until 1989 that a comprehensive standard was developed by the predecessor of the ITU, the Consultative Committee for International Telecommunications and Telegraphy (CCITT). The effort of the CCITT resulted in the Synchronous Digital Hierarchy (SDH) standard, which begins at a data transmission rate of 155.52 Mbits/s.

SDH was oriented as a transport mechanism for the European E carrier that was based on a format of thirty-two 64 Kbit/s DS0 channels. In North America the T carrier uses a DS1 signal consisting of 24 DS0 channels. Thus, the effort of the American National Standards Institute (ANSI) focused on the DS1 signal. Fortunately, there are many common features between the DS1 and E1 signals; thus, in an ANSI standard promulgated during 1997, SONET and SDH were defined as being compatible at rates of 155.2 Mbits/s and above. In this section we first discuss SONET, examining its line structure, frame format, and transmission hierarchy. Once this is accomplished, we will then examine SDH and conclude this section by focusing on common SONET and SDH topologies.

Overview

Both SONET and SDH represent optical technologies based on byte multiplexing. This means that each SONET or SDH node includes a time division multiplexer that is responsible for forming and demultiplexing the applicable optical signal. Although the cost of such equipment considerably declined during the late 1990s, it is still substantial. In addition, as we will note while reviewing the composition of SONET and SDH frames, their structure is quite complex and includes a considerable amount of overhead. Therefore, some carriers are beginning to bypass SONET and SDH and transport data directly over multiple optical channels created via WDM and DWDM. However, both SONET and SDH provide plenty of bang for the buck because they can support a ring topology, where a fiber cut can be almost instantaneously compensated for by rerouting of the optical signal in a reverse direction around the ring. This restore capability is critical for customers of many communications carriers since the Joe Backhoe effect could otherwise disrupt the ability of hundreds of thousands to millions of customers. Thus, we can expect both SONET and SDH to remain viable technologies for the foreseeable future.

The SONET Transmission Structure

SONET can be viewed as having a two-dimensional structure. In the horizontal dimension, SONET defines the method of data transmission between repeaters and other devices where an optical signal is transmitted and received. In the vertical dimension SONET defines a bit-rate hierarchy formed through the process of byte multiplexing.

Figure 6.4 illustrates the line structure of SONET. The SONET line structure was developed in recognition of the fact that an optical signal can flow through many intermediate devices on its path between source and destination. This resulted in the development of the layered line structure shown in Figure 6.4. Note that this structure enables the subdivision of the transmission path into distinct entities, thus facilitating SONET operation, administration, maintenance, and provisioning, a set of conditions referred to by the acronym OAM&P. Through this subdivision it becomes easier to refer to a portion of a path instead of a complete path, which facilitates such operations as performance measurements, testing, and troubleshooting on a segment basis. This, in turn, can expedite the isolation of problems to a specific area.

In examining Figure 6.4, note that the lowest layer of SONET is a section. A *section* in the SONET context represents a transmission path between two repeaters or other locations where the optical signal is transmitted and received. At the next-higher layer, the line represents a unit of SONET transmission capacity in terms of a *Synchronous Transport Signal* (STS), whose operating rates we will examine shortly. At the top of the line hierarchy is the *path,* which represents the end-to-end transmission of an STS payload by SONET. As we will note when we examine the composition of the STS frame format, a mechanism was created within frames to enable various OAM&P functions to be performed, such as providing a voice channel to be used by technicians.

The SONET Transmission Hierarchy

Under SONET the fundamental digital signal is referred to as the *Synchronous Transport Signal, level 1* (STS-1) and has an operating rate of 51.84 Mbits/s. Higher-level STS signals have bit rates that correspond to multiples of the fundamental STS-1 signal. Those higher-level signals are formed through the process of byte-interleaved multiplexing of lower-level signals.

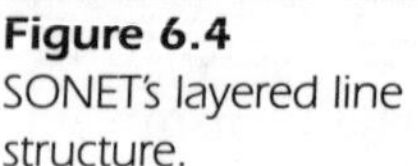

Figure 6.4
SONET's layered line structure.

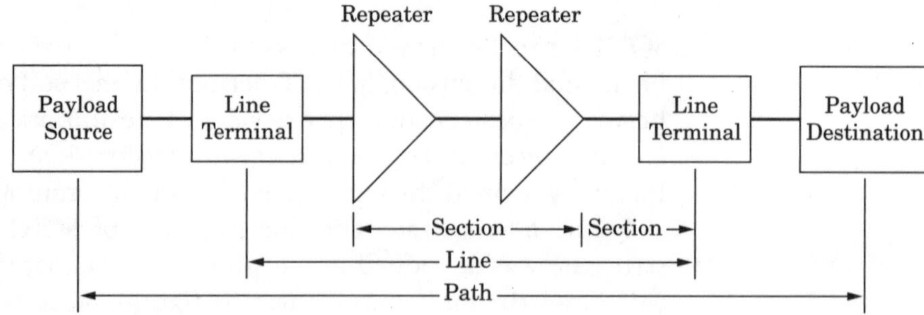

The actual STS signals refer to a sequence of bits and should not be considered as representing an electrical or optical signal. Thus, they more approximately represent an abstract signal. For each STS signal, there exists a corresponding optical transmission signal, referred to as an *optical carrier* (OC). The format of the OC signal is based on the corresponding STS signal format. The two lowest levels in the SONET hierarchy also have defined electrical signals: EC-1 and EC-3. Both of these signals facilitate the use of test instruments as well as the interconnection of transmission equipment.

Table 6.1 summarizes the SONET digital hierarchy. In examining the entries in Table 6.1, note that in trade publications it is quite common for a reference to STS-48 and STS-192 in the form of OC-48 and OC-192 to be rounded to data transmission rates of 2.5 and 10 Gbits/s, respectively. Similarly, OC-768, which is presently in field trials, is commonly referred to as a 40-Gbit/s *transmission link*.

If you periodically read trade literature, you may come across an OC term with a trailing "c," such as OC-192c. This term is usually employed to indicate a "clear channel" that provides a data transmission rate at the well-known optical signal rate.

THE STS-1 FRAME SONET's fundamental signal, STS-1, operates at 51.84 Mbits/s, whereas, as noted in Table 6.1, higher-level signals operate at 3× multiples of that rate.

Although the STS-1 frame represents a sequence of serially transmitted bits, most publications employ a two-dimensional matrix to illustrate the format of the frame. This two-dimensional matrix facilitates the

TABLE 6.1

The SONET Digital Hierarchy

Abstract signal	Buildup	Optical signal	Electrical signal	Bit rate, Mbits/s
STS-1	N/A	OC-1	EC-1	51.84
STS-3	3 × STS-1	OC-3	EC-3	155.52
STS-12	12 × STS-1	OC-12	N/A	622.08
STS-48	48 × STS-1	OC-48	N/A	2488.32
STS-192	192 × STS-1	OC-192	N/A	9953.28
STS-768	768 × STS-1	OC-768	N/A	39813.12

schematic representation of overhead byte locations and will also be used by this author. Thus, Figure 6.5 depicts the STS-1 frame as a two-dimensional series of rows and columns, also indicating the section and line overhead areas.

The omission of a path overhead area from Figure 6.5 is intentional. The rationale for this omission is that the path overhead area literally "floats" within the payload area. As we continue our examination of the STS-1 frame, we note how certain bytes within the section overhead area function as pointers to the actual location of a frame within the payload area shown in Figure 6.5. When illustrated in two dimensions, note that each row contains a series of 90 bytes, with the overhead repeating every 90 bytes.

The STS-1 frame is depicted in two dimensions as nine rows, each containing 90 bytes. This is because the STS-1 frame is 810 bytes in length. This frame repeats at the 8000 frame-per-second rate of both T1 and E1 carriers. Thus, 810 bytes/frame × 8000 frames/s results in a line operating rate of 51.84 Mbits/s.

In examining Figure 6.5, note that the first three columns in each row represent overhead. Thus, the actual gross payload of an STS-1 frame

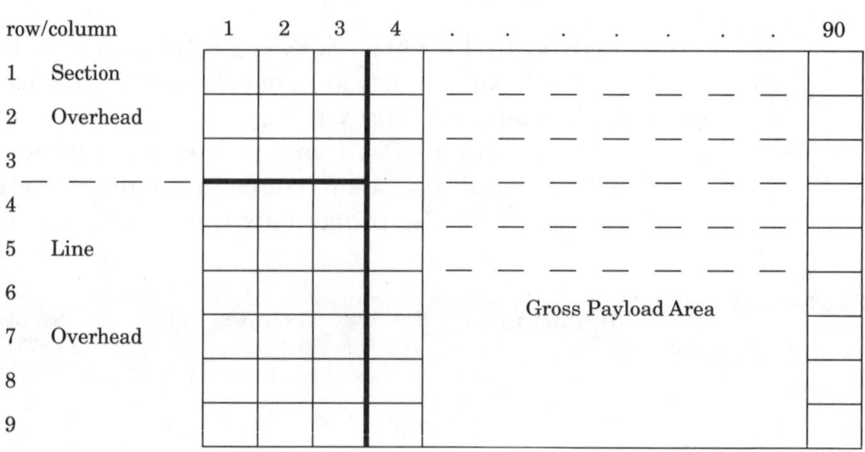

Figure 6.5 Two-dimensional depiction of the STS-1 frame.

becomes 783 bytes/frame × 8000 frames/s × 8 bits/byte, or 50.112 Mbits/s. As we will note below, there are nine path overhead bytes that float within the gross payload area. Thus, the net payload is 774 bytes/frame × 8000 frames/s × 8 bits/byte or 49.536 Mbits/s.

This net payload capacity is sufficient to transport a T3 or even an E3 payload. However, to do so SONET uses special overhead bytes, referred to as *payload pointers*, which indicate where the payload begins within the SONET frame. This use of pointers permits different types of payloads to literally float within the frame to an applicable position. Through the use of pointers individual DS1 frames can be synchronized, permitting SONET frames to be synchronized and providing for the "S" in the acronym SONET.

OVERHEAD As indicated in Fig. 6.5, the first three columns of each STS-1 frame are used for section layer and line layer overhead. The remaining 87 bytes in each row represent a gross payload of 783 bytes per STS-1 frame.

PAYLOAD The payload area of an STS-1 frame consists of 783 bytes. Although not shown in Fig. 6.5, the payload area is actually subdivided. The major portion of the payload area is used for the actual payload, while 9 bytes within the general payload area are used for path overhead. Thus, the net payload area available per STS-1 frame is actually 774 bytes.

The gross payload area is organized into a floating position within the STS-1 frame. Since the payload area floats, a mechanism is necessary to denote the actual position of the payload within the frame. This mechanism is provided by the use of two line overhead bytes: H1 and H2. In SONET terminology the total 783 bytes in the gross payload area, including 774 bytes of payload data and 9 bytes of path overhead data, is referred to as a *synchronous payload envelope* (SPE). The floating of the SPE provides a mechanism for overcoming differences in synchronization between different T- and E-carrier transmission facilities.

An example of a synchronous payload envelope floating within an STS-1 frame is shown in Fig. 6.6. Actually, the STS SPE can begin anywhere within an STS-1 envelope. Figure 6.6 shows an SPE beginning in one STS-1 frame and ending in the next frame. The line overhead bytes (H1 and H2) represent an offset in bytes within the gross payload area. The H1 and H2 bytes are allocated to a pointer that indicates the offset in bytes between the pointer and the first byte of the STS SPE. The H3 byte is used to compensate for clocking differences; that is, when a source clock is fast with respect to the STS-1 clock, the H3 byte can be used, permitting one payload byte from the SPE to be placed into the H3 byte. When this occurs, special coding is placed into byte locations H1 and H2.

As you examine the SPE positioning within the STS-1 frame shown in Fig. 6.6, note that the first path overhead byte has the designator J1. Because the payload wraps around to the next row, the second path overhead byte, designated B3, is aligned by column directly below the first path overhead byte. Similarly, the third path overhead byte is aligned under the second one, and so on.

OVERHEAD BYTES One of the key advantages of SONET and SDH as well as a key disadvantage of each is their structured hierarchy. The key advantages of the structured hierarchy facilitates OAM&P operations. Other advantages include the fact that an upgrade from one hierarchy to another as well as the multiplexing of lower layers into a higher layer are relatively simple time division multiplexing operations. Unfortunately, the price of these advantages is in the overhead associated with the need for section, line, and path overhead bytes. As we noted earlier, each STS-1 frame of 810 bytes has only 774 bytes available for the actual pay-

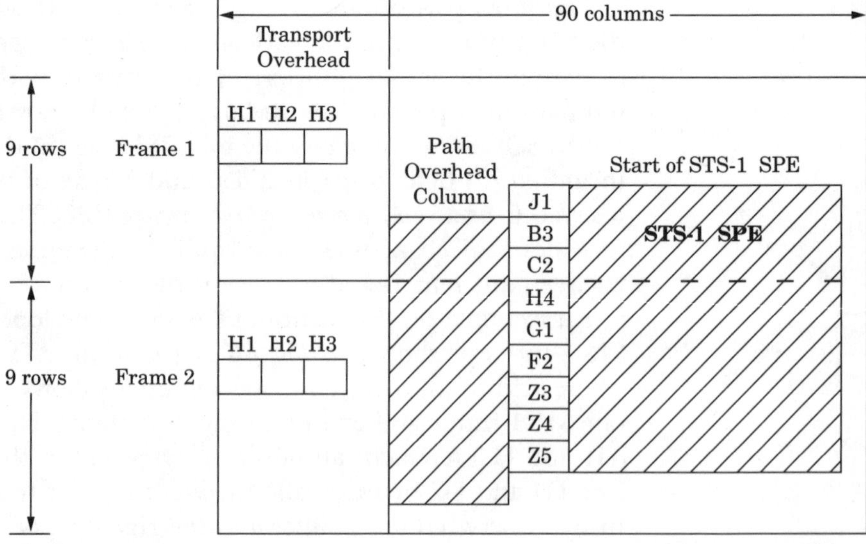

Figure 6.6 Positioning of the synchronous payload envelope (SPE) within an STS-1 frame.

load. For this reason, work commenced on transporting data directly over individual wavelengths provided by WDM and DWDM systems when this book was prepared. In the wonderful world of the Internet Protocol (IP), the technology being developed is referred to as *IP over DWDM.* Now that we have an appreciation for the advantages and disadvantages associated with the use of the three types of STS-1 overhead bytes, let's see how those bytes are used. In so doing, we will examine those bytes with respect to the area in which they are contained.

Section Overhead Bytes The term *section overhead bytes* refers to the path between repeaters or other locations where the optical signal is transmitted or received. Figure 6.7 illustrates the basic positioning of the STS-1 overhead bytes associated with an STS-1 payload. Note that 9 bytes represent section overhead, while 18 and 9 bytes represent line overhead and path overhead, respectively. Also note that while the section and line overhead bytes are in fixed positions, the path overhead can be aligned in any column other than the first three columns when the STS-1 frame is depicted as a 9 × 90 byte two-dimensional array.

In examining the overhead byte designators shown in Figure 6.7, note that rows 1 through 3 represent section overhead, resulting in 9 section overhead bytes. The A1 and A2 bytes are used to denote the beginning of an STS-1 frame. Thus, these 2 bytes are framing characters. A1 has the bit composition 11110110 or hex (hexadecimal) F6, while A2 has the bit

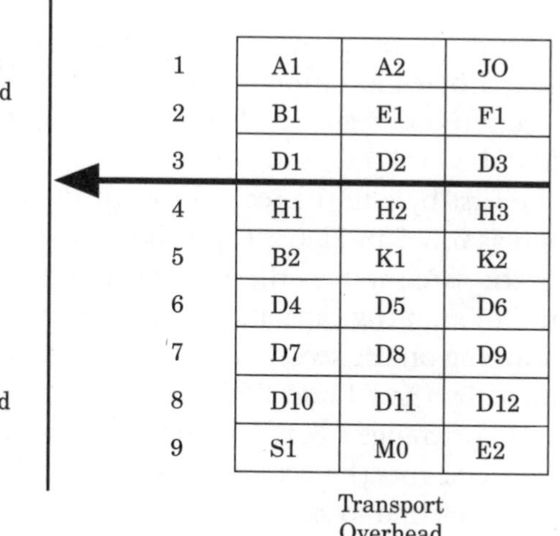

Figure 6.7 STS-1 overhead bytes.

composition 00101000 or hex 28. Because the A1 and A2 bytes appear horizontally in the two-dimensional matrix, they are also referred to collectively as the *horizontal sync pulse.*

The third byte in the first row in the section overhead is the J0 byte. The function of this byte is to verify that patching operations are performed correctly. Thus, the J0 byte represents a section-level trace byte.

The second row in the section overhead area commences with the B1 byte. The B1 byte is a bit interleaved parity code (BIP-8) byte. In an STS-1 frame even parity is used to check for transmission errors over a regenerator section, and its value is calculated over all bits of the previous frame. Thus, the B1 byte provides an error performance monitoring capability for a SONET section.

The second byte in the second row is the E1 byte. This byte represents a section-level order wire for voice communications between regenerators, huts, and remote terminal locations. The use of this byte provides a 64 kbit/s pulse-code modulation (PCM) voice channel between sections.

The third byte in the second row is the F1 byte. This byte represents a section user data channel, providing a 64-kbit/s path that can be read from or written to at each location where section terminating equipment is located.

The third row in the section overhead area contains three bytes labeled D1, D2, and D3. These bytes form a section data communications channel that operates at 192 kbits/s, which provides an OAM&P capability between section terminating equipment.

Line Overhead Bytes The line overhead consists of six rows, each containing 3 bytes, for a total of 18 bytes per STS-1 frame. The first row in the section overhead, which is row 4 in Figure 6.7, contains the previously described H1, H2, and H3 bytes.

The first byte in the second row of the line overhead section, which is shown as row 5 in Figure 6.7, is the B2 byte. This byte functions in essentially the same way as the B1 byte; however, it provides an error performance monitoring capability at the line layer instead of the section layer. Continuing on the second row of the line overhead section, the B2 byte is followed by the K1 and K2 bytes. The K1 and K2 bytes provide the capability to rearrange a SONET ring as well as to detect the alarm indication signal (AIS) and the remote detection indication (RDI) signal. These 2 bytes are referred to as the *automatic protection switching* (APS) bytes.

The third through fifth rows in the line overhead area, which equate to rows 6 through 8 in Figure 6.7, contain the bytes labeled D4 through D12. These nine bytes provide a 576-kbit/s data transmission rate for OAM&P use.

The last row in the line overhead area contains 3 bytes labeled S1, M0, and E1. The S1 byte represents a synchronization status byte. Four bits in this byte (bits 5 through 8) are used to transport the synchronization status of the network element. The following byte, which is labeled MO, represents a line remote error indication (REI) byte. If a receiver receives a corrupted signal, it will set the MO byte to indicate the occurrence of this situation. Last but not least, the E2 byte represents the last or final line overhead byte. This byte provides a voice channel that can be used by technicians and represents a 64-kbit/s order wire byte. As this is a voice PCM channel, this byte is ignored as it passes through regenerators.

Path Overhead Bytes In concluding our examination of STS-1 overhead bytes, we will consider path overhead bytes. Those bytes are primarily designed to convey information for the payload user, which is normally a communications carrier but could be an organization that installs a SONET facility within a campus or industrial complex. As noted in Figure 6.7, there are 9 bytes in an STS-1 path overhead signal.

The first byte, which is labeled J1, represents an STS path trace byte. This byte conveys a repetitively transmitted string, which enables a receiving terminal to verify that it is connected to the transmitting terminal.

The second path overhead byte is labeled B3. This byte provides a mechanism for path-level performance monitoring. To do so, the B3 byte contains a path BIP code.

The B3 byte is followed by the C2 byte. This byte indicates the type of payload or content of the STS SPE and is referred to as the *STS path signal label byte*. The fourth path overhead byte is labeled H4. This byte functions as a multiframe indicator and is meaningful for certain payloads. The fifth byte, which is labeled G1, is a path status byte. The function of this byte is to convey path error reporting back to the original path terminating equipment. The last three path overhead bytes are reserved for future use. As indicated in Figure 6.7, these bytes are labeled Z1 through Z3. Now that we have a general idea of the function of the three types of overhead bytes in an STS-1 frame, we will literally move up the SONET hierarchy and examine how higher SONET signals are created. Thus, let's study the format of an STS-N frame, where "N" refers to a multiplier of the 810 bytes of a basic STS-1 frame.

THE STS-N FRAME The STS-N frame represents a specific sequence of $N \times 810$ bytes formed by byte interleaving N STS-1 frames. Although the transport overhead bytes are aligned for each STS-1 frame before interleaving is performed, the associated SPEs do not have to be aligned. The rationale for this is the fact that the SPEs can float within an STS-1 frame, with their exact position denoted by the H1 and H2 pointers that were described earlier.

The first level of buildup of the STS-1 signal is obtained by byte interleaving of three such signals. This action results in the creation of an STS-3 frame, which can be viewed as a matrix of 9 rows and 270 columns. Similar to an STS-1 frame, an STS-3 frame contains section and line overhead bytes. While an engineer or communications carrier technician will need information about the function of those bytes and the placement of low-order signals, such as DS1, DS1C, and DS2s, within an STS-1 signal, we will not go that deep into the signal structure. Instead, we will observe the general structure of an STS-N frame, which is shown in Figure 6.8. Note that N can be 1, 3, 12, 48, 192, or even 768 and provides a two-dimensional structure such that the sequence of $N \times 3$ columns for 9 rows represents the transport overhead. Thus, when $N = 3$, there will be 9 columns (3×3) by 9 rows or a total of 81 available transport overhead bytes for section and line overhead. However, only a subset of the available transport overhead bytes are actually used in upper-hierarchy signals.

Now that we have an appreciation for SONET, let's discuss its international counterpart, SDH.

The SDH Hierarchy

As we noted earlier in this chapter, the Synchronous Digital Hierarchy (SDH) represents an international standard that is very similar to that of SONET. The key difference between the two resides in the fact that SONET begins with an STS-1 signal operating at 51.840 Mbits/s while SDH commences with an STM-1 (Synchronous Transport Module 1) signal which corresponds to the SONET STS-3 signal. In other words, SDH begins at 155.52 Mbits/s.

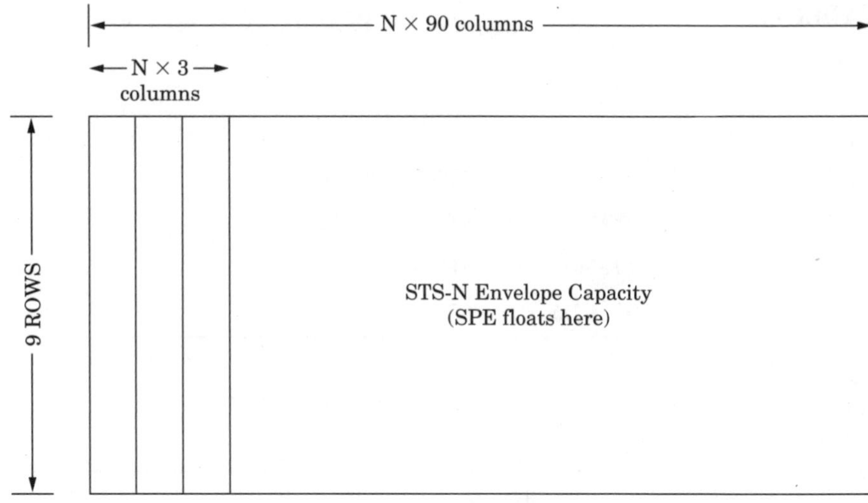

Figure 6.8 The general structure of the STS-N frame.

The SDH transmission hierarchy is listed in Table 6.2; however, one item in that table requires a bit of elaboration. That item is the STM-0 entry, since under SDH there is actually no STM-0 signal designation. However, because an STM-1 can be considered to consist of three STS-1 signals, we can use the designator STM-0 to refer to an STS-1 signal in the SDH hierarchy. Thus, SDH begins operations at 155.52 Mbits/s, which represents 3 times the lowest operating rate of SONET. However, the five levels of operation of SDH can be equated to the five higher layers in the SONET hierarchy. Another item that warrants a bit of elaboration is the STM-256 signal. Similar to a SONET's OC-768, this signal is commonly rounded and described as a 40 Gbit/s data transmission rate. Another difference between SONET and SDH that warrants attention is one of terminology. We can note this difference by examining the SDH frame.

THE STM-0 FRAME Figure 6.9 illustrates the basic structure of the SDH frame. If you compare Figure 6.9 to the basic structure of the STS-1 frame illustrated in Figure 6.5, you will note that they are nearly identical but employ different nomenclatures. For example, the repeater section is referred to as the *section overhead area* in an STS-1 frame. Another difference with respect to nomenclature involves the payload area. In an STM-0 frame the payload area is known as *administrative unit 3* (AU3). Since the payload floats, the H pointers used in the STM-1 frame were renamed as the AU3 pointer area.

TABLE 6.2
The SDH Transmission Hierarchy

Signal	Buildup	Operating rate, Mbits/s
STM-0	—	51.84
STM-1	3 × STM-0	155.52
STM-4	4 × STM-1	622.08
STM-16	16 × STM-1	2488.32
STM-64	64 × STM-1	9953.28
STM-256	256 × STM-1	39813.12

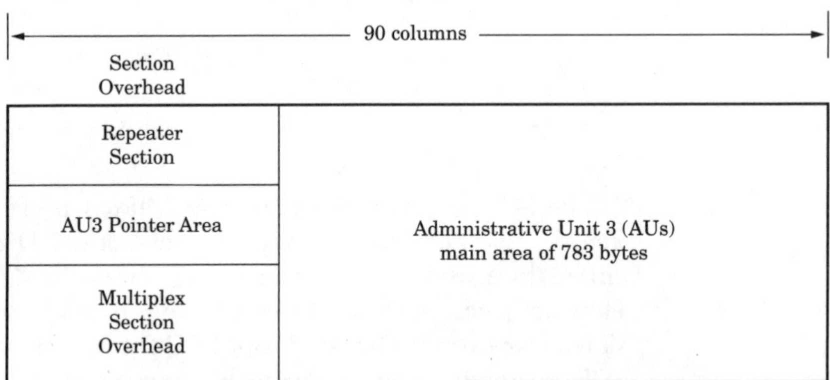

Figure 6.9 The STM-0 frame structure.

The STM-0 frame represents a virtual signal that forms an STM-1 frame. Once this SDH frame is formed, four STM-1 frames then form an STM-4 frame, and so on. Each STM frame represents an increase in the two-dimensional matrix used for representation similar to the manner in which an STS-N signal builds up. For example, an STM-1 frame would be represented as 9 rows times 270 columns, similar to an STS-3 frame, but using different technology. Now that we have a general idea of the SDH hierarchy, let's conclude this section by discussing some of the topologies supported by SONET and SDH.

Networking Topologies

Both SONET and SDH support a number of networking topologies. Those topologies range in scope from a simple point-to-point line structure to a self-healing ring architecture.

POINT-TO-POINT CONFIGURATION The most common and least costly network architecture supported by SONET and SDH is a point-to-point network configuration. This network configuration is commonly used to access a ring and also functions as a basic block for the construction of a ring. While a point-to-point configuration is more economical than a ring, many subscribers are now electing to access a carrier's communications infrastructure via a local ring since it provides a local loop fault-tolerance capability.

Figure 6.10 illustrates two versions of SONET and SDH point-to-point topologies. A basic point-to-point network configuration is shown at the top portion of the figure. Note that the path terminating terminal multiplexer (PTM) serves as a device which multiplexes multiple DS1s, E1s, and other signals onto an OC-1 signal in a SONET environment or onto the signal for an STM-0 frame in an SDH environment. As illustrated at the top of Figure 6.10, a basic point-to-point network configuration includes two PTMs connected via fiber, which may or may not require one or more optical repeaters or amplifiers.

POINT-TO-MULTIPOINT CONFIGURATION The basic point-to-point network structure shown at the top of Figure 6.10 can be easily modified into a point-to-multipoint structure. This modification is

Figure 6.10
Point-to-point and point-to-multipoint configurations.

a. Point-to-point

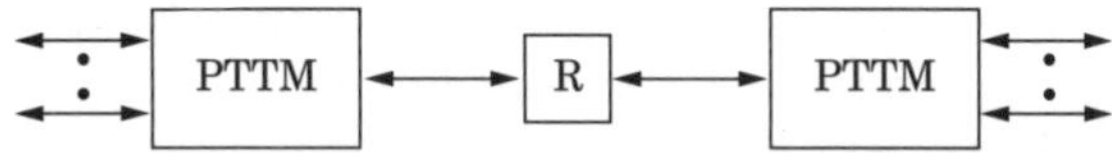

b. Point-to-multipoint

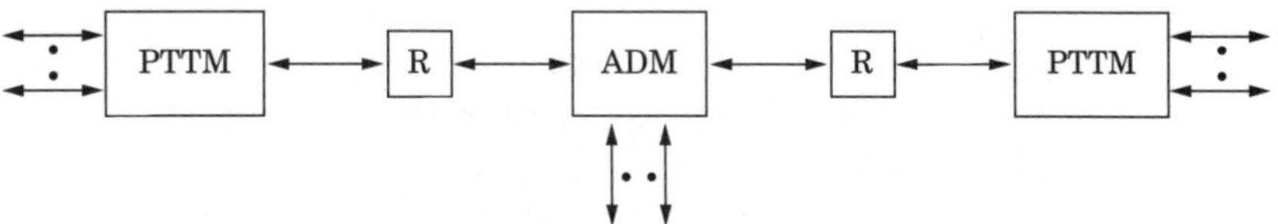

Legend:

PTTM	Path Terminating Terminal Multiplexer
ADM	Add/Drop Multiplexer
R	Repeater

accomplished through the insertion of an add/drop multiplexer (ADM) into the point-to-point network configuration. An example of a point-to-multipoint network configuration is shown in the lower portion of Figure 6.10. Note that the ADM can be a node on the network or can be routed to another multiplexer via a point-to-point network configuration. The advantage of the ADM is that it allows DS and E carrier signals to be added or removed without having to first demultiplex, then add or drop channels and then remultiplex data. This capability results from the fact that SONET and SDH frames include pointers that denote the location of signals within the frame. In addition to point-to-point and point-to-multipoint network configurations, both SONET SDH support hub and star network configurations. Thus, let's take a look at those two network configurations.

HUB A *hub* network configuration in a SONET or SDH environment is similar to that of a LAN hub, in which the hub represents the focal point or star where a collection of point-to-point connections terminate. However, unlike a LAN hub, a SONET or SDH hub employs a digital access and cross-connect system (DACCS) to route data between individual point-to-point lines. Figure 6.11 provides an example of a SONET or SDH hub-based network configuration. Because all data flow through the DACCS, this system provides the capability to monitor network performance and utilization from a single location. Because the DACCS functions as both a switch and an add/drop multiplexer, it is ideal for supporting unexpected network growth as well as affording the capability to restructure the configuration of a network to meet changing subscriber requirements.

Now that we are acquainted with the hub or star configuration, we will conclude our examination of SONET and SDH network configurations by turning our attention to the ring.

RING If you pick up a trade publication, you will probably encounter one or more communications carrier advertisements that discuss their ability to almost instantaneously compensate for an outage. What they are referring to is *survivability*, which was one of the design goals of SONET and SDH. To accomplish this goal, both technologies support a self-healing feature that enables a fiber cut to be immediately recognized and compensated for by switching traffic onto a different path formed by a ring structure. The alternate path is built into one design

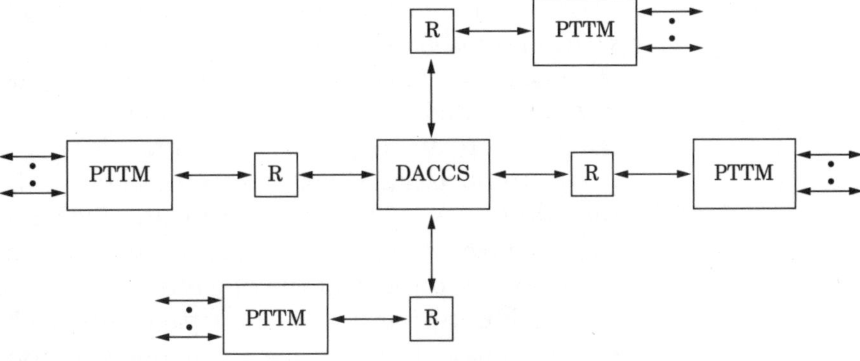

Figure 6.11 A hub or star network configuration.

for SONET and SDH. However, there are two types of rings: a *path-switched ring,* which switches individual paths, and a *line-switched ring,* which switches the entire capacity of the optical line. The key differences between the two ring topologies are the number of fibers used and the resulting cost of the ring. A path-switched ring uses two fibers and is less expensive to construct than a line-switched ring, which can use either two or four fibers. As we examine each type of ring, we will note that a line-switched ring can provide an additional level of switching capability over a path-switched ring.

PATH-SWITCHED RING A path-switched ring uses two fibers to form a concentric ring through the routing of point-to-point lines connecting individual nodes within a network. For each path between two nodes data are transmitted in both clockwise and counterclockwise directions, resulting in the term *bidirectional ring* used to refer to a path-switched ring configuration. Each receiving section in each node monitors both signals and selects the better one. Thus, any break in a signal between a pair of nodes allows the receiver to select the other fiber as a mechanism to continue operations.

LINE-SWITCHED RING In comparison to path switching, a line-switching topology provides the capability to switch paths at any node within a ring if a fiber failure occurs. Thus, a line-switching ring enables switching at either the line or the path level, which provides an additional level of switching capability in comparison to a path-switching ring. To accomplish this, one pair of fiber is idle but available for instantaneous utilization.

A second difference between path and line-sharing rings concerns the number of fibers supported by each topology. A path-sharing ring supports two fibers and simply selects the best signal. A line-switching ring supports the use of two and four fibers. When a two-fiber line-switched ring is employed, only a single pair is used to form the ring. In this configuration the capacity of the ring is partitioned into two equal portions, with each treated as if it were a separate line, and line-switching occurs on a path basis. However, when four fibers are used, it becomes possible to compensate for multiple fiber failures and switch on a line basis. Now that we understand the general topology supported by SONET and SDH, we will discuss WDM and DWDM operations in the WAN.

Wavelength Division Multiplexing (WDM) and Dense WDM (DWDM)

In the previous section we became acquainted with SONET and SDH operation. In that section we noted that both technologies use time division multiplexing on a byte interleaving basis to transport circuit-switched voice and leased-line services in the form of T- and E-carrier facilities.

Because SONET and SDH are limited to the static mapping of T- and E-carrier transmission facilities, they are not well suited for transporting bursty traffic, such as LAN-to-LAN transmission. For example, the connection of two 10BASE-T Ethernet LANs via a WAN may on average be satisfied by the use of a T1 line operating at 1.544 Mbits/s, which can be easily accommodated within a SONET or SDH transport. However, if the customer wanted the ability to burst transmission to 10 Mbits/s, the operator of the SONET or SDH facility would need to map the Ethernet signal to the next largest payload in the optical transport, which would be the 51.8-Mbit/s STS-1 frame in SONET.

To illustrate how inefficient SONET could become, let's assume that a communications carrier operating on a OC-12 ring had a significant number of requests from businesses to provide LAN-to-LAN connectivity over the ring. In this situation the OC-12 ring, which operates at approximately 622 Mbits/s, could support twelve 10-Mbit/s connections, in effect wasting 80 percent of the optical bandwidth.

If we consider emerging broadband service, such as the different flavors of digital subscriber lines (DSLs) and Cable Modem as well as the need of businesses to interconnect geographically separated locations and provide access to Web services, we can also note a growing problem. The frame structure and static mapping of SONET and SDH is inefficient for the transport of certain types of high-speed data services that cannot be easily multiplexed onto T1, T3, E1, and E3 circuits. Thus, the demand for bandwidth has increased not only because of an increase in data services but, in addition, because of the poor fit of certain types of data services into SONET and SDH frame structures. This growth in bandwidth demand is being satisfied in two ways. The most common method is the primary focus of this section, which is an increase in the capacity of fiber through wavelength division multiplexing (WDM) and dense WDM (DWDM). A second method, which is the focus of the end of this section, is the use of an alternative protocol stack.

WDM

Earlier in this chapter we briefly examined the rationale for WDM and three of the key technologies that made DWDM possible. In this section we expand our knowledge of both areas and familiarize ourselves with some specifics concerning WDM and its evolution into dense WDM.

RATIONALE The growth in the use of the Internet and graphical images by both commercial and residential telephone subscribers resulted in the demand behind the development of modern WDM systems. While just about every person on the planet is aware of the growth in the use of the Internet, the effect of graphical images on the infrastructure of communications carriers may not be as obvious as a driving force for WDM. Thus, let's take a brief look at the effect of graphics on transmission.

Until relatively recently the average email was approximately 2000 bytes in length. With the introduction of signature blocks and the placement of other graphics into an email, the size of this commonly used business communications tool considerably increased. For example, consider a small signature block rectangular in shape consisting of 180 × 64 pixels. When inserted into an email, the signature block by itself adds 180 × 64/8 bits per byte, or 1440 characters to the length of the email.

In both business and consumer environments, people are commonly using digital cameras for a variety of functions. When this author was involved in a fender-bender and took his car to the local insurance adjuster, the adjuster used a digital camera to photograph the modification made by a Land Rover bumping the rear of the author's sports car. The adjuster explained that damage reports for the insurance company's home office are now prepared using a word processing template with digital camera photographs attached and are then emailed to the company's home office. With the camera having a resolution of 1074 × 768 pixels, the resulting effect of adding just one picture has a pronounced effect on the transmission of an accident repair estimate report. For example, normally the insurance company's accident repair estimate report, prior to the use of a digital camera, was limited to hardcopy text and was approximately 4000 characters in length. Adding the equivalent of a 3 × 5-in photograph resulted in 1074 × 768 × 15 in, or 1.546 Mbytes if the photograph was in black and white. Of course, digital cameras take color photographs, so we would be remiss if we did not consider the effect of color. Because most digital cameras use "true color," which provides over 16 million colors using 24 bits associated with each pixel, to represent a distinct color, we need to multiply 1.546 Mbytes by 3. This is because it takes 3 bytes to represent 24-color bits. Doing so, we obtain both a data storage and data transmission requirement for 4.638 Mbytes of data. Although virtually all digital cameras use JPEG (Joint Photographic Experts Group) compression to reduce the storage requirements of pictures, in all probability the color photograph will be approximately 100 kbytes or more in size. If we compare that size to the 4-kbyte text-based report, we can note that just one photograph represents 25 times the storage and transmission requirements of the prior report. While this author could go on and provide many additional examples, like Indiana Jones, he will simply say "trust me" when referring to the fact that the explosion in the use of digital cameras, MP3 audio players, imbedded graphics in documents, and similar multimedia developments has resulted in a considerable increase in communications traffic, which became a driving force for WDM and its evolution into dense WDM.

EXPANDING CAPACITY Considering the accelerating growth in data transmission, it was a foregone conclusion that communications carriers needed additional capacity. When considering adding capacity to their infrastructure, communications carriers had four basic options:

1. When optical fiber was initially installed, most cables included a bundle of fibers, some of which were not immediately used and which are referred to as "dark fiber." Thus, one option that a communications carrier could consider was to illuminate existing dark fiber.

2. A communications carrier could also consider upgrading the transmission rate of existing optical fiber. For example, OC-48 operating at 2.5 Gbits/s could be upgraded to OC-192 operating at 10 Gbits/s. Unfortunately, upgrading the speed of an optical fiber is not a simple process. The ability to upgrade the transmission rate of a fiber depends first and foremost on the type of fiber. If the fiber was installed over the past few years, it is more than likely low-dispersion optical fiber that has the capability to operate at a higher data rate. If not, the fiber cannot support a higher transmission rate. If the existing optical fiber is capable of supporting a higher data rate, the fiber can be considered as a potential candidate for upgrading. The term "potential" is used because, even though a fiber can support a higher data rate, it may not be economically viable to replace existing equipment to include the laser and amplifiers in each hut, along with the route of the fiber and the photodetector at its termination location. Thus, a communications carrier will more than likely sharpen its pencil and consider other possibilities.

3. Another possibility is to install a completely new system to include the latest type of fiber that provides the potential to upgrade to OC-768's 40-Gbit/s transmission rate. However, as you might expect, this could be a very expensive solution to a capacity constraint.

4. Thus, another option, which has become the favorite of many communications carriers, is to transmit more lambdas (λ) or lightwaves per fiber through the use of WDM, DWDM, and, as the number of wavelengths exceeds 80, ultradense WDM, referred to as UDWDM.

Wavelength division multiplexing dates to the 1980s, when AT&T introduced its FT3C system. A wavelength division multiplexer operates by launching two or more optical signals at different wavelengths. The different wavelengths are coupled onto a common fiber, typically in the 1300- or 1550-nm region. When two wavelengths are coupled, this results in a doubling of the transmission capacity of a given optical fiber. During the early to mid-1990s several WDM systems were introduced that doubled to quadrupled the capacity of existing optical fibers.

OPERATION We can obtain an understanding of WDM operation by remembering our good friend who was introduced earlier in this book, Roy G. Biv. As we remember, Roy G. Biv represented a popular acronym for the different colors of light visible to the human eye: red, orange, yellow, green, blue, indigo, and violet. We also noted earlier in this book that while colors are transmitted through the air together, they can be easily separated by the use of a prism. Perhaps remembering Roy G. Biv and a high school physics class using a prism to separate light, early WDM systems injected light at 850 and 1300 nm into a multimode fiber using a simple fused coupler. At the distant end of the fiber, another coupler was employed to split the received light onto two short fibers. One fiber was connected to a silicon detector that was very sensitive to an 850-nm wavelength, while the second fiber employed a germanium or an InGaAs detector that was more sensitive to a 1300-nm wavelength. Through the use of optical filters unwanted wavelengths were removed, enabling each detector to receive the applicable wavelength that it was designed to receive.

Developments in optical technology, including the manufacture of economical single-mode fiber, resulted in most communication carriers installing this type of optical cable for use in their backbone infrastructure. At first, the primary method employed to multiplex multiple light signals onto a fiber was based on the use of couplers, allowing utilization of 1300- and 1550-nm optical windows where attenuation in terms of dB / km is minimized. Although light sources in the 1300- and 1550-nm optical windows were adequate to enable WDM, one problem that arose was that this technique was incompatible with the emerging design of fiber. Specifically, because of different dispersion characteristics, glass fiber was designed and manufactured differently for use in 1300- and 1550-nm systems. In fact, fiber that was optimized for operation in the 1300-nm optical window was used primarily for local metropolitan connections while the backbone infrastructure of communications carriers, including long-haul and submarine cables, were based on the use of dispersion-shifted optical fiber that was optimized for performance in the 1550-nm optical window.

The development of optical amplifiers during the late 1980s was based on the efforts of researchers at Bell Laboratories, then part of AT&T and NTT of Japan. As we noted earlier in this chapter, erbium amplifiers were developed by doping a length of fiber and then pumping the fiber at a wavelength of 980 nm. The laser pump provides the energy for the amplifier while the incoming signal results in stimulated emissions as light pulses pass through the section of the doped fiber. Because the stimulated emission results in the stimulation of additional emissions, the effect is a rapid growth in the flow of photons in the doped portion of the fiber. Figure 6.12 illustrates the components of an erbium optical amplifier. Note that the laser pump, coupler, and a small section of erbium-doped fiber are typically assembled in a common housing. The effect of the erbium amplifier can result in a gain of approximately 40 dB, or 10,000 times the received signal.

The use of couplers made it relatively easy to increase the number of light sources being combined onto a common fiber. A more difficult task facing engineers was the demultiplexing of the common light source. To separate the different wavelengths coupled onto a fiber, developers used a mirrorlike device known as a *grating* that works in essentially the same way as a prism. Specifically, the grating is used to separate light sources into their individual wavelengths by reflecting them at different angels. Figure 6.13 illustrates the wavelength division multiplexing process, which may be a tongue twister but that actually represents the process of distributing the different wavelengths to their intended destinations. Note that the grating must be extremely accurate in reflecting different wavelengths to applicable output fibers.

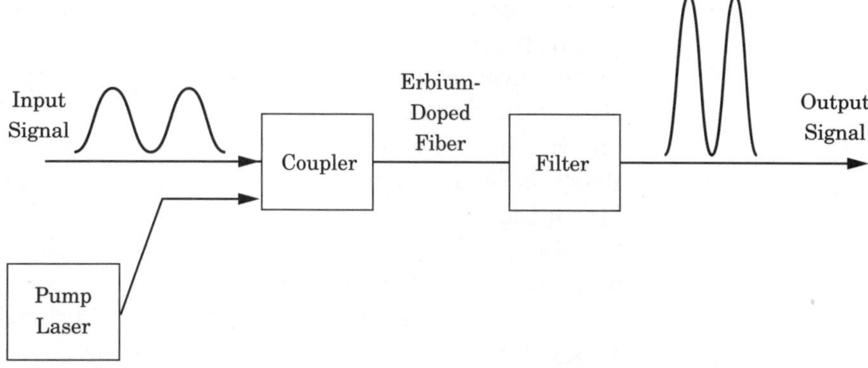

Figure 6.12 An erbium amplifier consists of a pump laser, a coupler, and a small section of erbium-doped fiber.

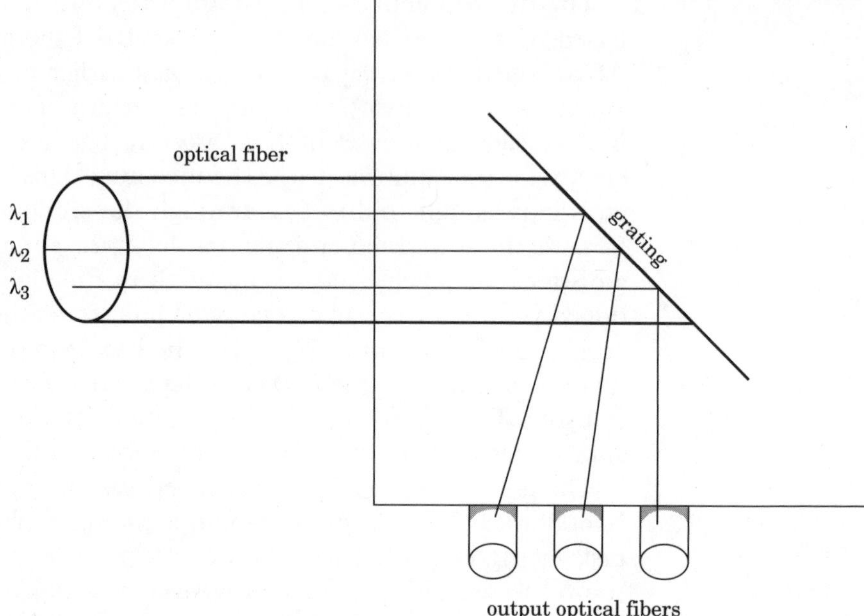

Figure 6.13 WDM demultiplexing.

The actual gratings shown in Figure 6.13 are referred to as *Bragg gratings* or *in-fiber Bragg gratings.* The use of this type of grating results in a wavelength-dependent reflector that can reflect different wavelengths at different angles. The actual grating consists of a length of optical fiber in which the refractive index of the core is modified. As a result, a portion of the modified fiber acts as a narrow-bandwidth filter whose function is similar to that of a mirror, reflecting specified wavelengths. Using fiber grating, it becomes possible to reflect each wavelength on the common fiber. Thus, if we refer back to Figure 6.13, we can note that the grating results in the reflection of each of the three individual wavelengths transported on the fiber.

One problem associated with increasing the capacity of existing fiber plant is the manner in which existing SONET and SDH systems operate. The vast majority of existing communications carrier fiber plant is single-mode fiber, which has a high dispersion rate in the 1550-nm window. This made it rather difficult to obtain an OC-192 or STM-64 transmission for long distances and resulted in a majority of upgrades stopping at the OC-48 2.4-Gbit/s operating rate. Thus, communications carriers were faced with the choice of installing additional fiber or deploying denser WDM if they wanted to continue to increase their bandwidth.

Needless to say, various vendors, including Ciena, Lucent, and Nortel, increased the capacity of their WDM equipment to accommodate the requirements of communications carriers, resulting in the development of DWDM equipment.

DWDM

Dense wavelength division multiplexing (DWDM) is a technique that enables more optical wavelengths to be independently transmitted on a single fiber, with each wavelength operating at the maximum rate permitted by the fiber plant. To accomplish this, DWDM uses smaller spacing between wavelengths, such as 0.8 nm for a 16-wavelength system and 0.4 nm for a 32-wavelength system. Although there is no clear division between the point where WDM becomes DWDM, ITU standards stipulate DWDM 0.8-nm separations, so an increase in capacity of 16 times or more is representative of a DWDM system. However, some publications consider anything in excess of four derived optical channels to represent a DWDM system.

OPERATION The most common utilization of DWDM systems is to provide additional capacity to existing OC-48 systems. This is because the jump from OC-48 to OC-192 requires either some form of dispersion compensating fiber or new fiber using non-zero dispersion-shifted fiber (NZDSF), which costs approximately 50 percent more than traditional single-mode fiber. Because the vast majority of fiber plant installed during the 1980s and 1990s was single-mode fiber, the ability to operate at OC-192 or STM-64 is limited primarily to relatively new communications carriers that in the late 1990s and through 2001 installed nationwide fiber systems.

The most common form of DWDM uses a fiber pair, where one optical fiber is used for transmission while the other fiber functions as a receiver. Figure 6.14 illustrates the use of a 16-channel DWDM system. In examining this figure, it was assumed that amplifiers were required every 200 km and that the span distance between locations was 1000 km.

CAPACITY If we assume that the original optical pair shown in Figure 6.14 operated at 2.5 Gbits/s, then the use of a 16-channel DWDM system would increase the capacity of the system to 2.5 Gbits/s × 16 or 40 Gbits/s. In addition, if separate fiber pairs were installed instead of DWDM equipment, the communications carrier would require 4 × 16

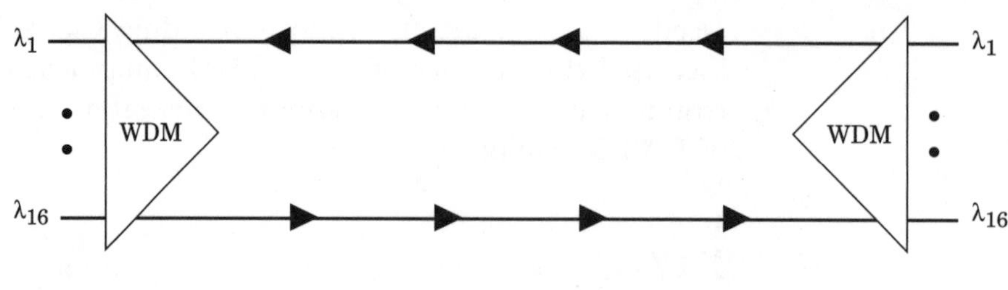

Figure 6.14 A 16-channel DWDM system.

or 64 amplifiers instead of 4. If the carrier simply bit the bullet and installed all new fiber, such as NZDSF which can support OC-192 at 10 Gbits/s, the higher operating rate requires optical regenerators at approximately every 40 km. Thus, although four fiber pairs could support the 40 Gbit/s rate obtained from DWDM, the number of regenerators would grow to 100.

WDM and DWDM Developments

In much the same way as the evolution of a 1908 baseball play was referred to as "Tinker to Evers to Chance," which referred to the key infield players on the Cubs, we can use the acronyms WDM to DWDM to UDWDM to denote the evolution of wavelength division multiplexing. During the late 1990s systems with 32 to 64 channels operating at OC-48 data transmission rates resulted in the ability of a single optical fiber to support a composite transmission rate between 32 × 2.5 Gbits/s or 80 Gbits/s and 64 × 2.5 Gbits/s or 160 Gbits/s.

Because OC-48 operating rates permitted the use of a wide range of optical fiber, most WDM systems, which were soon referred to as *DWDM*, operated their individual wavelengths at 2.5 Gbits/s. However, by the new millennium, the low-dispersion fiber being installed by many relatively recently formed telecommunications companies permitted OC-192 at 10 Gbits/s to become a reality. In competition, well-established communications carriers added new optical fiber to their infrastructure, which enabled them to upgrade to OC-192.

Although migration from OC-48 to OC-192 may appear to permit a composite increase by a factor of 4 in the data transmission rate capability of a fiber, the actual gain is more complex. This is because in an OC-48-based DWDM system signals are separated from one another by 0.4 nm (50 GHz), but in OC-192 signals are commonly allocated 0.8-nm slots. Thus, upgrading to an OC-192 DWDM system could result in a quadrupling of the transmission rate on half the number of channels, resulting in a doubling of the composite rate carried by an optical fiber. However, before we put down our pencils, a few words about evolving systems are in order.

Nortel has introduced an OC-192 based DWDM system with 0.4-nm signal separation. In so doing, Nortel transmits alternating wavelength channels in opposite directions to minimize the effect of optical crosstalk. Here optical crosstalk is similar to its electrical counterpart, representing the leakage of one optical channel into another. Figure 6.15 illustrates an example of optical crosstalk.

In examining Figure 6.15, you can note one of the problems facing DWDM developers as they attempt to add more channels to produce UDWDM systems. Specifically, as the spacing between channels narrows, crosstalk will increase. One method being examined to circumvent narrowing channel space is to use new wavelengths, such as those in the L band. Thus, work on 128-, 256-, and even 1024-channel DWDM systems that are referred to by many people in the industry as UDWDM are based on the use of wavelengths in the L band.

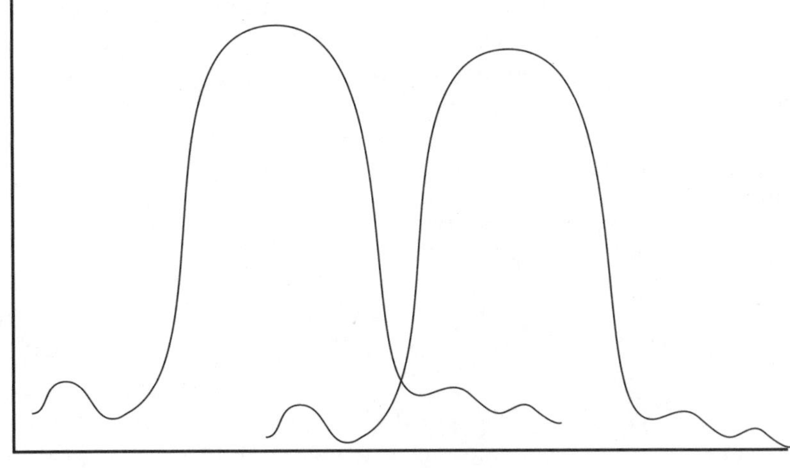

Figure 6.15
Optical crosstalk represents the leakage of one optical channel into another.

Alternative Protocol Stacks

Now that we have an appreciation for WDM and DWDM, we will conclude our discussion of fiber in the WAN by briefly examining the emergence of alternative protocol stacks to direct SONET and SDH in the WAN that can result in a more efficient transmission capability. Those alternative protocol stacks include ATM over SONET/SDH, IP over SONET/SDH, and IP directly over fiber. Figure 6.16 illustrates these three approaches to obtaining an enhanced transport capability.

The first alternative protocol stack shown in Figure 6.16, ATM over SONET/SDH, adapts all services to ATM cells, which are then transported via an existing SONET/SDH frame structure. Because ATM provides true quality of service (QoS) and represents a cell-based delivery service, it can better handle bursty traffic from LAN internetworking requirements as well as T- and E-carrier circuit emulation. In fact, Telcordia, formerly known as Bellcore, developed several standards for ATM over SONET/SDH. Two of those standards are GR-2837, which defines ATM virtual path support on SONET rings; and GR-2842, which denotes the requirements for ATM-based access multiplexers.

A second approach to obtaining an increased level of efficiency is indicated by the second protocol stack shown in Figure 6.16. In this example IP frames are transported over SONET/SDH networks. Currently there are no standards governing the transport of IP frames over SONET/SDH. However, IP could be used within traditional T- and E-carrier facilities to be carried in fixed positions within a SONET or SDH frame.

The third approach shown in Figure 6.16 is to adapt all services to IP frames for direct transport over an optical network, bypassing the use

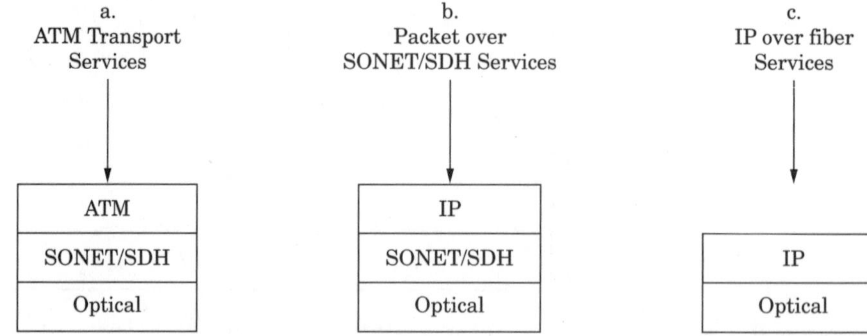

Figure 6.16 Alternate emerging optical transport facilities.

of SONET/SDH. Although there are no standards for this approach, some relatively newly formed communications carriers have implemented this approach using DWDM to obtain multiple IP transports at the OC-48 and OC-192 rates, thus providing a tremendous bandwidth capacity as well as the efficiencies of IP frames being transmitted only when there are data to transmit. Of the three approaches, it is the educated guess of this author that IP over fiber will evolve as the method of choice for many carriers.

CHAPTER 7

Fiber in the Neighborhood

The objective of this chapter is to provide readers with information concerning two potential large and evolving markets for fiber-optic components. One market, usually referred to as *fiber to the curb* (FTTC), represents the extension of fiber-optic technology to a central location within a residential or business area. From the curb, transmission occurs via conventional copper cable into the home or office. The second market for the use of fiber-optic components that we will examine in this chapter is *fiber to the home* (FTTH), which can also be considered to represent *fiber to the office* (FTTO), although this author is not aware of the second term currently being used. However, rather than discuss the routing of fiber-optic cable directly into an office building under the term *fiber to the home*, this author will use the term *fiber to the office* to represent the routing of optical fiber into office buildings. The title of this chapter stems from the fact that the routing of fiber to the curb, home, or office normally occurs from a central point within a neighborhood.

In this chapter we will first focus on the current and evolving local infrastructure of the two primary operators of communications services in the neighborhood: telephone companies and cable TV (CATV) operators. In so doing, we will note that both local telephone companies and CATV operators are changing their network infrastructure by routing fiber into neighborhoods, which can include fiber to the curb. However, because of the high cost associated with rewiring hundreds of millions of existing last-drop twisted-pair and coaxial-cable connections, it is doubtful that fiber to the home will become anything other than a limited networking strategy. Recognizing this, we will conclude this chapter by describing and discussing where fiber to the home makes economic sense.

The Existing Telephone Company Infrastructure

This section focuses on the existing telephone company network infrastructure at the local level. We will note how subscribers are serviced by a central or end office and why modern broadband communications support requires a modification to the existing telephone company infrastructure at the local level.

Overview

The telephone company represents a technology that is over 100 years old. During this period of time the basic infrastructure of most telephone companies at the local level has remained relatively stable. At the local level of the telephone company a central or end office is used to interconnect subscribers within a limited geographic area. That area typically has a 3- or 4-mile (approximately 15,000- to 20,000-ft) radius. Figure 7.1 illustrates the typical service area of a telephone company central office.

The mileage support radius is based on the transmission distance of an analog signal over copper twisted-pair wire. That radius varies because different wire gauge can be used to connect a telephone company end office or central office to each subscriber. Because a thicker wire has less resistance to the flow of electrons, it is possible to extend the radius of support from a central office by using thick wire. To avoid any confusion, it should be noted that wire gauge in the United States is expressed in an inverse relationship based on the American Wire Gauge (AWG) standard; that is, the greater the AWG number, the thinner the wire. Thus, a lower AWG number reflects a thicker wire that has less resistance and extends the radius of support from a central office.

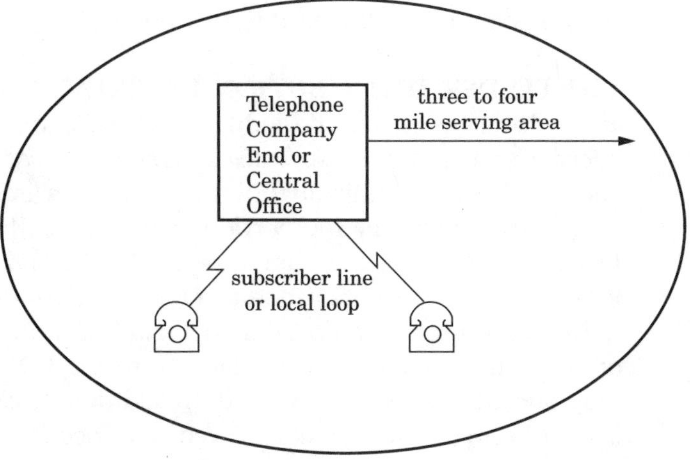

Figure 7.1 The typical service area of a telephone company central office has a radius of 3 to 4 miles from the office.

The Central Office

As a phone call is routed through one or more telephone networks, it will eventually flow to a central office. That office directly serves subscribers and routes calls to each subscriber via a copper twisted-pair wire. Because the central office is the last or end office in the hierarchy of telephone company offices, it is also commonly referred to as an *end office.* The twisted-pair wire installed from the end office to each subscriber is referred to as a *subscriber line* or *local loop.*

Each central or end office originally supported an individual three-digit exchange prefix, such as 477. Because four digits follow the prefix, this scheme resulted in an end office servicing a maximum of 10,000 subscribers. With advances in technology, switches installed in central offices in metropolitan locations may serve subscribers with 10 or more telephone number prefixes. However, the basic structure of the local loop has not changed.

Connection Methods

Because of economics, the local loop originating from a subscriber may not directly flow on an individual twisted pair to a telephone company central office. Instead, subscriber lines are typically routed to a neighborhood location where a number of subscriber lines are either bundled together for routing to the serving telephone company central office or are multiplexed to that office.

THE FEEDER DISTRIBUTION INTERFACE Bundling that occurs at the location where individual wire pairs are grouped together is referred to as a *feeder distribution interface* (FDI). At the FDI, up to 1000 and possibly more individual twisted-pair wires are bundled and routed to a central office. The top portion of Figure 7.2 illustrates the use of an FDI to interconnect a grouping of residential and business subscribers located within a neighborhood to a telephone company central office.

In examining the top portion of Figure 7.2, note that through the use of a bundle of copper pairs routed from the FDI to the central office, only one right-of-way is required. In addition, because a common path is used between the FDI and the central office, it is easier and less costly to route the bundled copper.

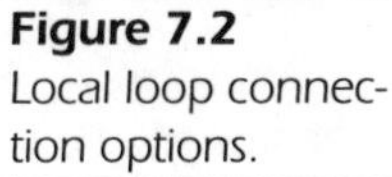

Figure 7.2
Local loop connection options.

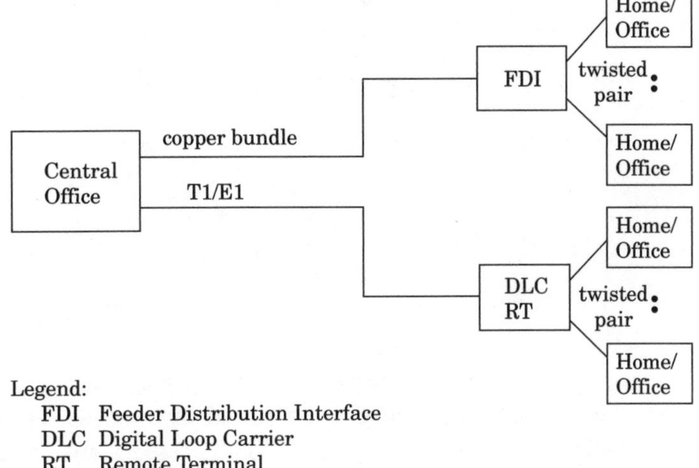

THE DIGITAL LOOP CARRIER REMOTE TERMINAL A second method commonly used to connect a large number of subscribers to a central office involves the use of a copper-based T1 or E1 transmission lines. In this method subscriber lines are first routed to a *digital loop carrier* (DLC), which represents an intermediate location between a group of subscribers and the central office. In telephone company terminology, these intermediate locations are referred to as *remote terminals* (RTs). Thus, the second method used to connect a large number of subscribers to a central office is referred to as *digital loop carrier remote terminal* (DLC RT), which is illustrated in the lower portion of Figure 7.2. Depending on the type of copper-based line routed to the DLC RT, either 24 or 30 voice channels can be supported. The T1 line supports 24 voice channels, while the E1 supports 30.

The use of a DLC considerably reduces the cost associated with supporting subscribers since it enables groups of 24 or 30 subscribers to be serviced via a common T1 or E1 transmission facility. Unfortunately, the use of a DLC has a major disadvantage. To understand this disadvantage, we must briefly review the method by which T1 and E1 transmission facilities operate. That method is based on time division multiplexing.

LIMITATIONS OF TIME DIVISION MULTIPLEXING (TDM) Figure 7.3 illustrates the use of a T1 multiplexer to transport 24-voice conversations over a common twisted-pair wire. In examining this figure,

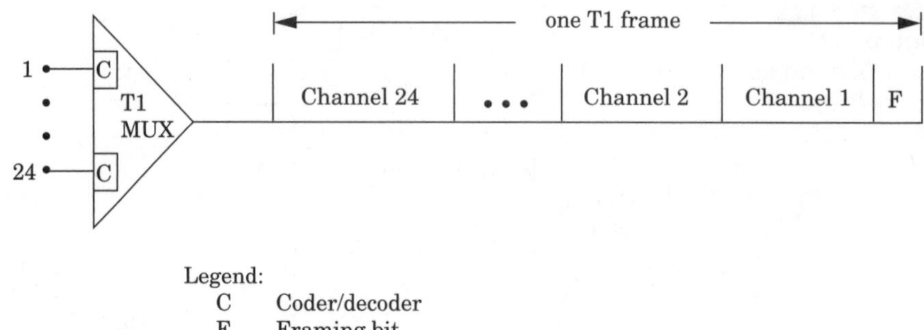

Figure 7.3 Using a T1 multiplexer.

note that the *coder* portion of the coder/decoder represents a module that converts an analog voice signal into a pulse-code modulation (PCM) digitized voice signal. The PCM digitized signal results from samples of the analog voice signal occurring 8000 times per second and each sample being converted into an 8-bit value. Thus, the data transmission rate of each digitized voice conversation represents a 64-kbit/s data stream. In the opposite direction, the *decoder* portion of the coder/decoder converts the digital signal back into its equivalent analog signal.

In examining the use of the T1 multiplexer shown in Figure 7.3, note that 24 voice inputs are digitized. A basic T1 frame consists of one 8-bit sample from each of the 24 voice channels plus a 1-bit framing bit that provides synchronization between T1 multiplexers. Thus, the operating rate of the T1 circuit becomes 193 bits/frame $\times$ 8000 frames/s, or 1.544 Mbits/s.

In an era of increasing use of broadband communications, such as digital subscriber lines (DSLs), the use of T1 and E1 circuits to interconnect a central office to a DLC RT located within a neighborhood represents a problem. That problem is one of capacity. A DSL typically provides a transmission rate of a minimum of 640 kbits/s in the downstream direction from the telephone company central office to the subscriber, while transmission in the uplink direction from the subscriber to the telephone company normally occurs at a minimum rate of 160 kbits/s. To obtain this transmission capacity, a DSL modem modulates data at frequencies beyond the 0- to 4-kHz range used by human voice.

Today telephone companies support over 10 types of DSL products, with most operating by splitting the bandwidth of the twisted-pair wire into two distinct channels beyond the 0 to 4 kHz used to transport human voice. Figure 7.4 illustrates the general frequency allocation on the local loop to the subscriber for one type of DSL service referred to as *asymmetrical digital subscriber line* (ADSL). The name of this DSL version indicates one key characteristic of its operation, namely, that the data transfer is asymmetrical.

In examining Figure 7.4, note that when ADSL operates on a subscriber line, there are actually three channels formed by frequency. One channel, which represents approximately 4 kHz of bandwidth, is allocated for the existing voice telephone operations. A second segment of frequency, which begins at approximately 30 kHz, is allocated to an upstream data channel, while a third segment of frequency is allocated to a downstream data channel. The third segment of frequency is much wider than the second segment to provide a downstream operating rate that is higher than the upstream operating rate. The reason for this asymmetrical design is that for Web surfing, transmission upstream is primarily in the form of relatively short Uniform Resource Locator (URL) Web page addresses, while transmission in the downstream direction consists primarily of Web pages. Because the number of bytes required to transport graphical images and text on a Web page considerably exceeds the number of bytes in a URL, bandwidth for the downstream data channel considerably exceeds the bandwidth allocated for the upstream channel. This results in a higher data transmission rate being obtainable on the downstream channel, which should satisfy the operational requirements of a vast majority of Web surfers.

If we turn our focus back to Figure 7.3, we will note that the T1 multiplexer uses a fixed format for allocating digitized voice channels onto a T1 line. That allocation results in twenty-four 64-kbit/s time slots becoming available for use. In this modern era of broadband communications where a single DSL-capable subscriber line can use all or most

Figure 7.4
ADSL frequency allocation.

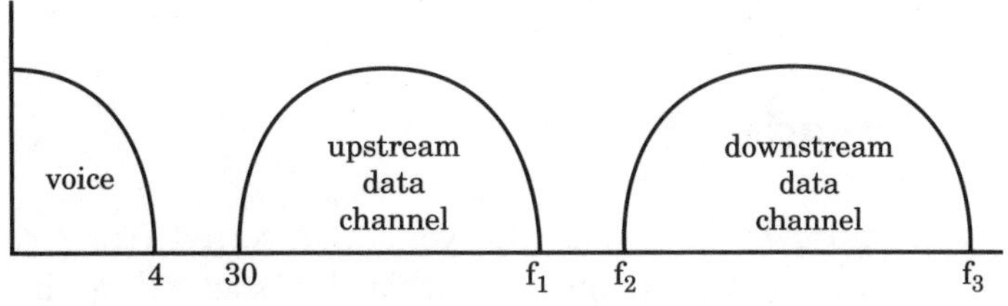

of the capacity of a T1 or E1 transmission line, the use of a DLC RT is insufficient and represents a bottleneck. Thus, the existing infrastructure of local telephone companies must be rebuilt to support broadband communications, a topic that we will focus on later in this chapter. Now that we have an appreciation for the local telephone company infrastructure that provides connectivity to subscribers via local loops, let's turn our attention to the cable television (CATV) infrastructure.

The Original Cable TV Infrastructure

In this section we discuss the original cable TV infrastructure, which remained essentially unchanged for almost 40 years. We will also learn how information flows on the original cable TV network infrastructure and why that infrastructure requires modification to support modern two-way communications used by cable modems.

Overview

Cable TV was originally developed as a mechanism to provide unidirectional video signals to subscribers. To accomplish this task, the wiring infrastructure was designed in the general format of an inverted tree. Thus, the head of the structure, referred to as the *headend,* resembles the root of a tree and emits signals that flow onto branches either directly to subscribers or which are further split onto additional branches, also known as *subbranches,* for routing to subscribers. Depending on the distance from the headend to subscribers, amplifiers may be installed on main branches and subbranches.

Headend

Figure 7.5 illustrates the general inverted tree structure of a cable TV transmission system. Note that the headend of the system is commonly connected to one or more satellite earth stations. The earth stations are

Figure 7.5
The basic cable TV network structure.

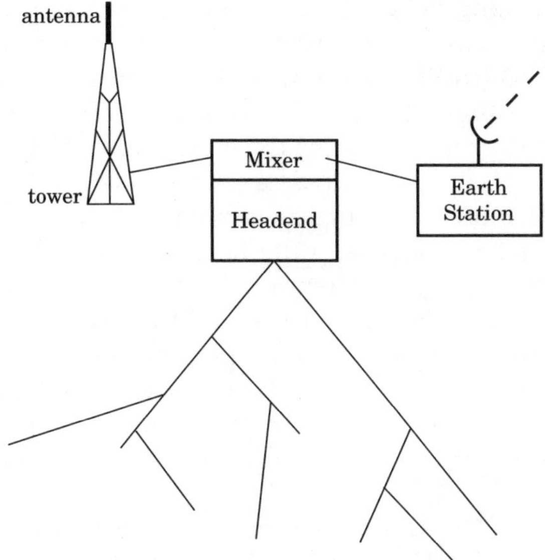

The basic cable TV network structure resembles an inverted tree.

receive-only (RO) satellite dishes designed to receive television signals transmitted via one or more satellites. Each satellite has a series of transponders that receive and rebroadcast TV signals; thus, you can consider the satellite transponder to represent a big audiovideo relay station in the sky.

To ensure that the fixed location earth stations receive the TV signals throughout the day, geostationary satellites are used. Such satellites orbit the earth at a distance of approximately 26,000 miles above ground level, enabling the satellite to rotate in tandem with the rotation of the earth. In this manner the satellite stays in a fixed location relative to the ground, beaming such programming as HBO (Home Box Office), Showtime, the nightly news of major television networks, and other programs that many readers watch each day.

SATELLITE RECEIVERS At the cable system headend, satellite receivers in the form of antennas mounted on an earth station or a nearby tower receive local programming from the immediate area. Both satellite-received programming and local programming are modulated by frequency onto the cable TV operator's network, with specific programming placed on predefined channels. The end result of this procedure is to enable cable TV subscribers to—either manually or via a remote controller—tune their CATV boxes to specific channels to watch desired programs. Of course, premium programming will be encrypted prior to entering the mixer shown in Figure 7.5.

THE MIXER The mixer takes input signals from different sources and combines them onto the headend. In so doing, the mixer places each received signal from a satellite or antenna onto a predefined channel for broadcast onto the CATV distribution system. In actual operation, the mixer receives input at different frequencies and moves such input to other predefined frequencies. Because the predefined frequencies represent channels, we can simply state that the mixer moves input signals to their appropriate channel placements.

AMPLIFIER Another device physically located within the headend is an amplifier. The amplifier boosts the level of the received signal so that subscribers can receive a high-quality signal. As we continue our exploration of the original CATV infrastructure, we will note that the headend amplifier is one of many located within the cable network.

In addition to an amplifier located at the headend, there are other amplifiers located throughout the CATV operator's serving area. Each amplifier boosts signal power, and its power-in/power-out ratio is referred to as its *gain.* Amplifier gain is measured in decibel milliwatts (dBmW). Recognizing the physical phenomenon in which high frequencies attenuate more rapidly than do low frequencies, CATV amplifiers are designed to provide a higher level of gain at high frequencies. If you are familiar with the manner in which high-speed modems automatically generate attenuation equalization, you will note that this nonuniform level of amplifier gain increases in tandem with frequency.

The Cable TV Distribution System

Figure 7.6 illustrates the original CATV distribution network. Note that the amplifiers are unidirectional as the original CATV network design was based on the one-way broadcast of programming to subscribers. It is well known that the ability to use a cable modem requires a bidirectional transmission capability. This means that the original CATV network infrastructure based on the use of unidirectional amplifiers was

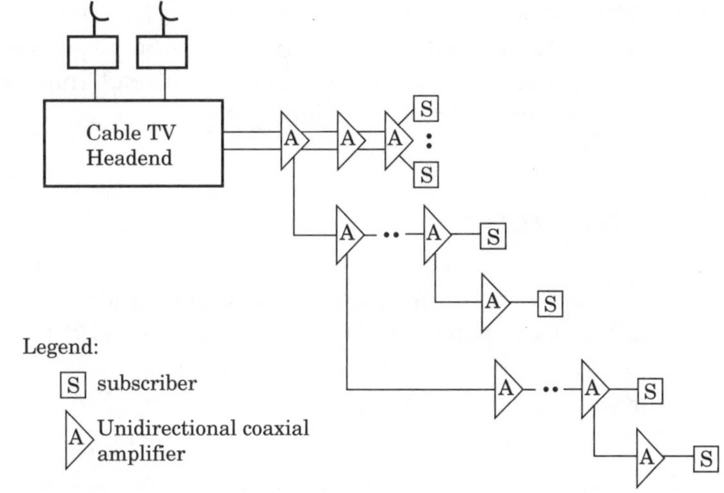

Figure 7.6 The initial cable TV network infrastructure was based on the use of one-way amplifiers.

incapable of supporting the use of cable modems. As an interim measure, some CATV operators initially marketed cable modems that used the public switched telephone network (PSTN) for uplink communications, while downlink communications occurred via the coaxial-cable infrastructure. Because this represents an interim solution as CATV operators invest billions of dollars in upgrading their infrastructure to support bidirectional data services, we will focus on the use of bidirectional transmission over the CATV infrastructure in the remainder of this chapter.

Now that we have a general idea of how the original CATV network infrastructure was designed, let's focus on the evolving telephone and CATV network infrastructure. In so doing, we will note the routing of fiber to the neighborhood, including the practicality of fiber to the curb and fiber to the home.

The Evolving Local Telephone Network

In this section we will examine the evolving local telephone network. Here the key term to note is "local," since the use of fiber in the backbone of long-distance communications carriers commenced during the

1970s. In comparison, the use of fiber in the local telephone network is a fairly recent phenomenon, since it is more expensive to replace cable in urban areas than place fiber on the railroad and pipeline rights-of-way commonly used to link cities together.

Overview

The ability to support broadband communication in the local telephone company depends on the type of cabling used to link subscribers to a central office and the distance from the subscriber to the central office. Basically, one of three methods can be used to connect telephone company subscribers to the serving central office:

1. Subscribers can be directly cabled to the central office. However, this method would be very expensive and is seldom used.
2. Customers can be cabled to a central office by routing subscriber lines to a common location within a neighborhood where they are bundled together for routing via a common path to a central office. While this procedure maintains the integrity of point-to-point routing from subscriber to a central office, it requires only one right-of-way for a majority of the cabling distance.
3. Cabling between subscribers and a central office can also be effected via multiplexing subscriber conversations into applicable time slots at a common location where a *digital loop carrier remote terminal* (DLC RT) is located.

Now that we have an appreciation for the three connection methods, let's look at them and their problems, if any, in some detail.

DIRECT CABLE The use of a direct cable connection between a subscriber and a central office can support broadband communications in the form of DSL use as long as the subscriber is located close enough to the central office. If the subscriber is located beyond an 18,000-ft radius from the central office, the ability to obtain a high level of transmission capability using DSL becomes problematic.

BUNDLING The use of a *feeder distribution interface* (FDI) involves the bundling of cables from each subscriber at a common point for routing to the central office. Because subscriber lines are not directly routed to the central office, this adds additional cabling distance and may render some DSL service problematic. In addition, when FDI support is utilized, the higher frequencies used by DSL modems can potentially create interference among the individual twisted-pair wires in the bundle.

DSL For reasons already mentioned, the use of a DLC RT network structure is not capable of supporting DSL transmission. As a brief refresher, the DLC RT multiplexer uses fixed 64-kbit/s time slots for each subscriber as it was developed during an era prior to emergence of the need for broadband communications. Now that the need for subscribers to transmit data beyond the 64-kbit/s capacity of the time slot allocated to each subscriber has arisen, the DLC RT is becoming obsolete.

Options

For the reasons mentioned above, many telephone companies are installing fiber from their central offices to neighborhoods. At those neighborhood locations a splitter can be used on each twisted-pair termination to separate voice and data channels. Voice can be amplified and placed back into existing copper bundles or the DLC RT infrastructure, while DSL modem data can be aggregated via the use of a *digital subscriber line access multiplexer* (DSLAM) for transmission via a fiber-optic cable to the central office, where it again bypasses the normal voice network.

Another option being used by some telephone companies involves replacing their existing copper links into a neighborhood with optical fiber that is then used for both voice and data communications. In this networking environment a multiplexer is used to digitize each voice channel and to aggregate both voice and data from subscribers for transmission to the central office, where voice and data are separated from one another.

Figure 7.7 illustrates one example of the emerging telephone company wiring infrastructure. In this example the multiplexers located within each neighborhood are shown supporting both voice and data from the neighborhood to the central office.

In examining Figure 7.7, note that the fiber trunk replaces a relatively long "backhaul" use of twisted-pair copper wire. Because DSL operations

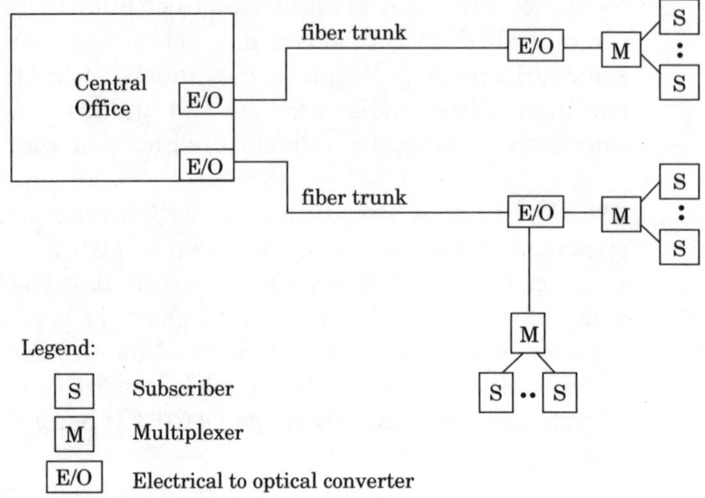

Figure 7.7 An example of the emerging telephone company wiring infrastructure that uses fiber to the neighborhood.

are distance-sensitive, the use of fiber provides telephone companies with a mechanism to increase the transmission rate that subscribers can obtain. In addition, the routing of fiber permits neighborhoods that would be normally too distant from a central office for DSL services via copper twisted-pair cabling to obtain the ability to be eligible for DSL services.

The Evolving Cable TV Infrastructure

In this section we discuss the evolving CATV infrastructure. We will also note how the development of the Internet and its access via cable modems required a major revision of 45-year-old (at the time of writing) network architecture.

Overview

Although early cable television systems were developed during the late 1940s as a mechanism to improve reception in rural areas, through the mid-1990s the architecture remained relatively stable. Commencing in the mid-1990s, the growth in the use of the Internet and its graphics-intensive Web pages provided the foundation for the development of a variety of broadband access techniques, including cable modems.

In comparison to the telephone network infrastructure, where DSL represents a nonshared, point-to-point technology, the transmission of data over a CATV infrastructure occurs on a shared-network basis. In this arrangement, one or more branches in the network may be configured to support the shared use of two TV channels for the transmission of data on a branch or series of branches to and from groups of subscribers. Thus, one channel is used for transmission in the downstream direction while the second channel is used for transmission in the uplink direction.

Similar to most flavors of DSL, the operation of a cable modem represents an asymmetrical transmission method. In the upstream direction, a data transmission rate between 640 kbits/s and 10.24 Mbits/s is supported in accordance with the method of modulation used as specified by a standard referred to as *Data Over Cable System Interface Specification* (DOCSIS). In contrast, in the downstream direction a data rate of either 27 or 36 Mbits/s is supported.

Bandwidth Limitations

While a bandwidth of 27 or 36 Mbits/s may appear quite sufficient for most applications, a cable modem subscriber shares that bandwidth with other subscribers. Thus, if the cable modem operator is greedy and assigns a channel to 1000 active subscribers, then the average effective bandwidth per subscriber is not 36 Mbits/s but 36 Mbits/s/1000, or 36,000 bits/s, which could be less than that of a conventional modem. For this reason, some cable TV operators limit a serving neighborhood for cable modem service to 500 subscribers.

Because the use of two bands of frequency on a CATV network must be limited to a neighborhood of 500 homes or offices before performance is significantly affected, the architecture of the CATV network must be revised. This revision requires much more than the simple replacement of unidirectional amplifiers with bidirectional amplifiers and explains why cable TV operators have dug deeply into their pockets to spend tens of billions of dollars upgrading their systems' infrastructure. One method used to upgrade a cable network involves the replacement of an all-coaxial-cable-based network with a hybrid fiber-coaxial system.

Hybrid Fiber-Coaxial Systems

The employment of hybrid fiber-coaxial (HFC) systems by CATV operators commenced during the mid-1990s in response to the need to provide not only support for more channels but also the ability to aggregate and route data originating from groups of subscribers to the headend and from the headend to a group of subscribers. In the downstream direction, routers must be placed at appropriate locations within the network infrastructure to send data to the applicable branch or subbranch, where the subscriber is located.

Figure 7.8 illustrates the emerging hybrid fiber-coaxial (coax) network that many cable TV systems are migrating to. Under the HFC network architecture a star cabling infrastructure is employed, where fiber-optic cable is routed from a location in a neighborhood referred to as a *fiber node* (FN) back to the CATV headend.

In examining Figure 7.8, note that in addition to an electrical to optical converter, each fiber node that supports data transmission to one or more branches will have a router that will be used to aggregate transmission to and from the branch feeder cables. The router's transmission would be placed on a separate wavelength for transmission in the uplink direction toward the cable TV headend. In the downlink direction a separate wavelength could be used to provide a data path to each router. However, programming was initially transmitted in the downlink direction via time division multiplexing (TDM). Thus, many cable operators continue to use TDM to each fiber node, where channels in each TDM slot are converted from the time domain into the frequency domain. The latter is required since subscribers tune their set-top boxes via frequency to locate a particular program of interest.

RATIONALE The rationale for the use of fiber into a neighborhood is based on economics, capacity, and fiber-optic immunity to electromagnetic interference. Because bandwidth capacity of fiber is several orders of magnitude above that of coaxial cable, the existing coaxial-cable drops can be used as is. Thus, the use of fiber provides the cable

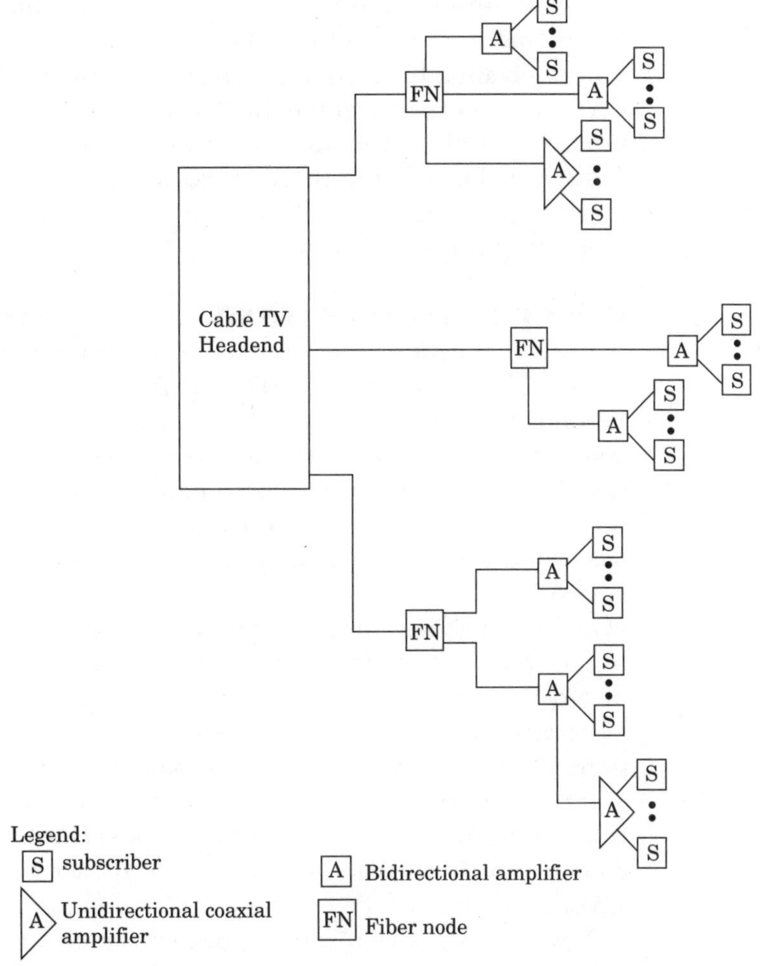

Figure 7.8
A hybrid fiber-coax distribution system permits the gradual upgrade of coax drop to support bidirectional transmission and also increases the number of channels of programming that can be carried by the CATV system.

operator with an economical method for upgrading their distribution system.

AMPLIFIER CONSIDERATIONS To ensure that the upgraded system is capable of supporting two-way communications, the unidirectional amplifiers within the remaining coaxial portion of the network must be replaced by bidirectional amplifiers. This then permits upstream transmission from subscribers to be supported.

In examining Figure 7.8, note that several amplifiers are shown as being unidirectional. While subscribers literally "at the end of the coax" cannot obtain a bidirectional transmission capability, this indicates how CATV operators can gradually "build out" their new infrastructure instead of having to replace everything at once. This also explains why subscribers located in different areas within a city or in different rural areas within close proximity to one another may not be able to use a cable modem at the same time.

CAPACITY CONSIDERATIONS Another advantage associated with the use of a hybrid fiber-coax (HFC) system is the additional capacity made available. Under the HFC architecture, downstream frequencies beyond the typical 550-MHz limit of a fully coaxial system becomes available for use. If we remember our physics, high frequencies on an electrical system attenuate more rapidly than do low frequencies. This phenomenon would require cable operators to place attenuation equalizers within their systems at closer spacings, which would drive up their network cost. Putting pen to paper, cable operators decided that it would be more advantageous to limit transmission to 550 MHz, resulting in most systems initially installed limiting support to 42 channels of programming.

Because the use of optical fiber provides a minimal loss of signal strength into a neighborhood, it becomes possible to support additional channels over coaxial cable that now has a shorter run length from the fiber node into homes and offices. Thus, HFC systems can support downstream frequencies of 750 MHz or higher, permitting support for 125 or more channels of programming.

Now that we have a general idea of the direction where the CATV infrastructure is headed, we conclude this chapter by focusing on a limited use of fiber. That limited use we will call *fiber to the home,* which at the beginning of this chapter we described as including fiber routed directly to residential houses as well as apartment buildings and offices.

Fiber to the Home

As mentioned at the beginning of this chapter and reviewed at the end of the previous section, we use the term *fiber to the home* to refer to the installation of optical fiber by a communications carrier to a specific building, such as a house or an apartment or office building. In this section we first briefly examine the history of fiber to the home and the reason why its deployment to date has been "underwhelming." Once this is accomplished, we will focus on the one area where fiber to the home represents a practical technology. This area concerns the provision of large buildings and their occupants with the ability to bypass the local telephone company, and, as you might expect, has economics as its driving force.

Overview

Fiber to the home represents a technology that service providers have experimented with since the 1980s. In spite of several well-publicized trials, FTTH represents a technology which, according to the most optimistic forecasts, is not expected to achieve a penetration level of more than 1 percent of all U.S. households until the year 2005 or so.

Fiber to the home (FTTH) represents a technology with a minimum market penetration rate primarily because of economics and competition with other broadband solutions that retain the ability to use the existing twisted-pair and coaxial-cable wiring that flows into several hundred million homes on a worldwide basis.

Economics

The use of fiber into a home or office that is already constructed requires new cabling to the building. Because of the cost associated with the laying of individual fibers to each building, most FTTH equipment placed in field trials uses splices on a fiber strand to serve four or more homes. One example of such technology is the "deep fiber" equipment from Marconi Communications, which can serve up to four homes with three telephone lines and provides a 25-Mbit/s data channel, analog TV, and a direct broadcast satellite (DBS) digital TV signal. At the time this book was being written Marconi was testing its equipment with both Bell South and Verizon Communications, with the latter representing the company formed by the merger of Bell Atlantic and GTE Corporation.

While using a single fiber to serve multiple homes reduces costs, the need to splice subbranches flowing to individual homes adds to the cost of the installation. In addition, fiber to the home must compete with the various flavors of digital subscriber lines that telephone companies and competitive local exchange carriers (CLECs) are offering, as well as cable modems being offered by CATV operators. Thus, it is difficult to be competitive when one vendor needs to install totally new wiring in the form of fiber to the home while a competitor may be able to offer the services required by a subscriber by simply mailing the subscriber a cable modem or DSL modem for self-installation.

Probably the best market for fiber to the home is when a bundled service including telephone, data, and video can be marketed and sold to subscribers. Perhaps recognizing this, Lucent Technologies has demonstrated equipment capable of simultaneously supporting four telephone calls, high-quality digital video, and data transmission at 82 Mbits/s. Lucent equipment requires a dedicated fiber to serve each home, which may not be more costly than using a fiber with splicing to multiple homes. However, Lucent equipment is similar to other optical fiber equipment in that the optical fiber reflects a need for new cabling for all homes other than those under construction or in the planning stage.

Since several million new homes and apartments are constructed each year in the United States, it may be possible for a communications carrier to work with builders to offer home buyers the ability to select a bundled service option based on the cabling of optical fiber to the home while the home is under construction. Now that we have an appreciation for some of the competitive issues that make it difficult for fiber to the home to be economically justified, let's examine the one area where business is literally booming. That area is known as a *bypass* technology.

Bypass

The mention of the term *bypass* to a local telephone company is in many ways similar to a matador waving a red flag at a bull. The reason for the ill-will toward this term is the fact that it results in a loss of revenue to the communications carrier. There are many types of bypass operations, ranging from the manner in which a residential customer bypasses the long-distance telephone company by dialing a number using a "1010" prefix, to the apartment or office building manager, which permits a competitive local exchange carrier to directly connect their building to the CLEC network. In this section we will focus on the latter, as it is normally accomplished via the use of optical fiber.

Figure 7.9 illustrates the relationship between cabling performed by a CLEC and an existing local exchange carrier (LEC); the latter term is used to denote the incumbent telephone company. Although the CLEC may be independently located, it can also have its office within the incumbent telephone company central office. Thus, the primary difference between the two concerns the type and structure of cabling.

The incumbent telephone company built their network infrastructure over a period of approximately 50 to 100 years by routing copper cable directly into buildings. In comparison, a CLEC providing a bypass capability that enables subscribers to avoid the high cost of local terminations commonly constructs a fiber loop in an urban area with termination points at buildings that sign up for their services. One of the earliest firms to offer this type of service was Metropolitan

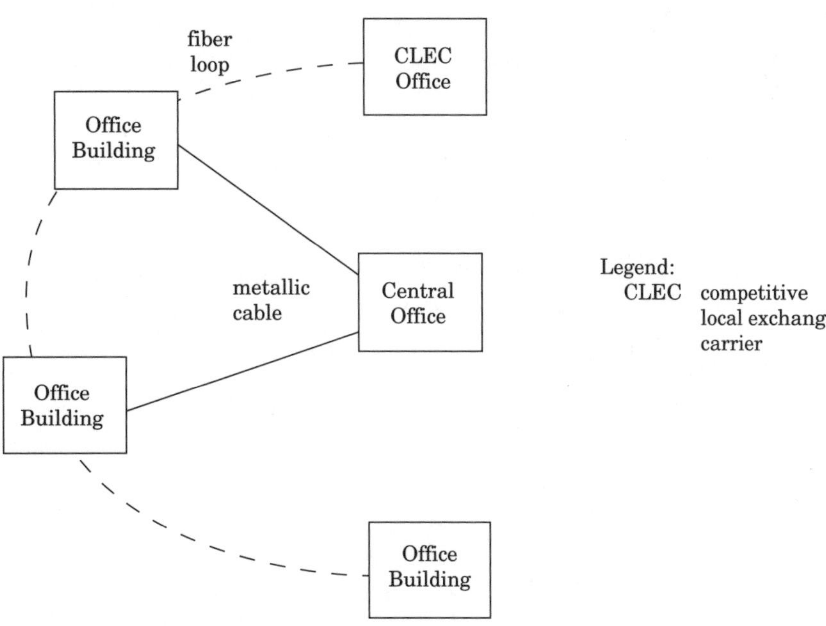

Figure 7.9 Comparison of cabling methods of the incumbent telephone company and a competitive local exchange carrier.

Fiber Services (MFS), which is now part of Worldcom Communications. Today there are literally over 100 CLECs providing a bypass of the local telephone company; however, because it is expensive to provide connectivity to a building, the primary focus is on installing a fiber loop that provides connectivity to office buildings that may have hundreds to thousands of workers. While the derogatory term "cherry picking" is used to infer that the bypass provides access only to larger customers in terms of billing, it also explains why bypass "bypasses" smaller buildings and homes where the cost of providing service would be higher than any expected profit from providing the service. Unlike the local telephone company that must provide universal service, there is no such requirement for bypass operators. Thus, they are free to select large revenue sources and bypass the small consumer, which explains why FTTH including fiber being routed into office buildings and apartment buildings is not expected to significantly increase its penetration rate before the year 2006 or so.

CHAPTER 8

Fiber in the Building

Until this chapter our primary focus concerning optical equipment was either on distinct components, such as lasers, LEDs, and photodetectors, or devices used in a LAN or WAN environment. In this chapter we focus on fiber in the building, examining the operation and utilization of such products as the fiber modem, fiber multiplexer, and even an optical mode converter. Although a local area network is obviously a network facility installed primarily within a building, this topic is not covered in this chapter. Instead, LANs were covered as a separate entity in Chapter 5 when we examined fiber in the LAN. Thus, the primary focus of this chapter is on the use of optical fiber and equipment used within a building primarily to obtain a transmission capacity extending beyond that provided by conventional copper cable.

Fiber-Optic Modems

Any communications system requires a transmitter, a transmission medium, and a receiver. In an optical transmission system we normally and correctly consider a light-emitting diode (LED) or laser to represent the transmitter. However, when an LED and photodetector are packaged in a common housing to provide a product that converts a serial data stream flowing over copper cable into an optical signal and converts the optical signal into an electrical signal in the reverse direction; manufacturers refer to this device as a *fiber modem* or *fiber-optic modem*. Thus, in this section we will examine the different types of fiber-optic modems and how they can be utilized for intra- and even interbuilding communications.

Basic Operation

The objective behind the use of a fiber-optic modem is to extend the transmission distance between two devices with electrical interfaces. For example, when cabling a terminal on one floor of a building to a controller located on another floor, it is often a distance limitation that precludes installation of the terminal where one desires it to be located. Instead of being forced to compromise on the location of the terminal, the use of a pair of fiber-optic modems and a pair of optical fibers can often provide a considerable degree of flexibility. Similarly, consider the location where a T1 or E1 circuit enters a building, often referred to as the *demarcation* or *DEMARC point*. Because the width of a digital pulse is inversely proportional to the data transmission rate, this means that high-speed transmission such as via T1 and E1 circuits has relatively narrow pulse widths. This also means that such pulses can travel only a limited distance before requiring the use of repeaters to regenerate the digital signal. This also explains why most communications carriers place the DEMARC point at a location close to the main wiring closet or PBX (private branch exchange) room. Doing so permits the organization to comply with the wiring distance limit between the DEMARC and equipment that terminates the T1 or E1 circuit, such as a channel service unit. Unfortunately, very rarely do the communications requirements of an organization remain static. Thus, requirements can be expected to vary over time to include the need for additional T1 or E1 lines that must be connected to equipment that will be installed at other locations within a building.

Because it is relatively expensive for the communications carrier to change the routing of an underground or aboveground main cable into a building, they will more than likely continue to route new circuits to a common DEMARC point, even if the customer would prefer another location. Thus, once again the use of a pair of fiber-optic modems and optical fiber may represent the solution to obtaining the connectivity required within a building.

Employment Features

Figure 8.1 illustrates the basic employment of a pair of fiber-optic modems to extend the transmission distance between two copper interfaces. In examining Figure 8.1, note that the type of copper interface built into the fiber-optic modem is not specified as this is one of the features that separates one type of fiber-optic modem from another. Also note that the optical fiber is specified as either multimode or single-mode. While some optical modems are capable of supporting either type of fiber, other devices are restricted for use with a specific type of device. Thus, the type of optical fiber supported by the optical modem can be an important selection criterion.

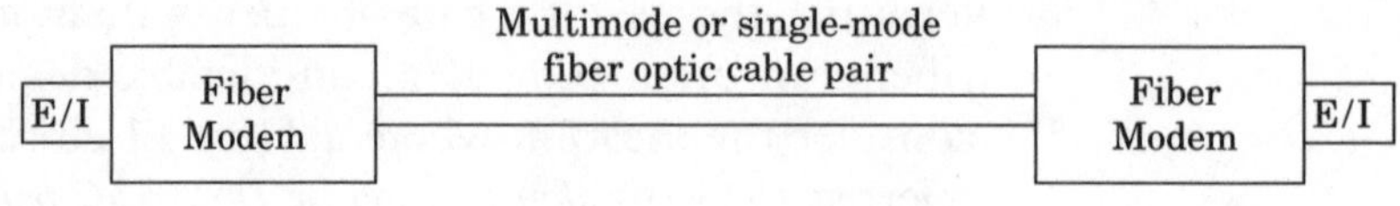

Figure 8.1 Using a pair of fiber modems to extend the transmission distance between devices with electrical interfaces.

Because LEDs require a larger core than do lasers, the type of transmitter—LED or laser—will normally govern the type of fiber supported. Other important optical modem selection criteria include the data transmission rate supported at the electric interface, the type of data, asynchronous or synchronous supported at the interface, and the optical loss budget, which will determine the transmission distance obtainable between the pair of optical modems.

Returning to Figure 8.1, it is also important to note that the copper interface can be a single wire pair (a two-wire), a pair of two-wires (a four-wire), or such connectors as RS-232 that have 25 conductors or another type of connector. Thus, many parameters must be considered when selecting a fiber modem.

Selection Features

To facilitate the selection of an appropriate fiber-optic modem, Table 8.1 lists some of the modem products being marketed by TC Communications of Irvine, California at the time this book was prepared. If you examine the entries in Table 8.1, you will note that TC Communications provides a fiber-optic modem that can support extension of the transmission distance of just about every type of electrical interface used in a communications environment. Such popular EIA electrical interfaces as RS-232, RS-442, RS-485, and RS-530, as well as T1, T3, and V.35 connectors, are supported by different fiber-optic modems manufactured by TC Communications as well as other vendors. Because of the wide range of products from different vendors, we will focus on TC Communications products to see what optical modem selection factors you may wish to consider and what options are available. In examining optical modem selection factors, we will examine the TC Communications Model TC1540 fiber-optic modem, which is illustrated in Figure 8.2.

TABLE 8.1

Representative Fiber-Optic Modems

Transmission mode	Electrical interface	Electrical connector	Loss budget, dB	DTE/ DCE* switch	Maximum data transmission rate
Asynchronous	RS-232	DB25F	12	Yes	19.2 kbits/s
Asynchronous/ synchronous	RS-232	DB25F	15	DCE	125 kbits/s
Asynchronous	RS-232	DB25F	15	DCE	125 kbits/s
Asynchronous	RS-232	DB25F	15	DCE	125 kbits/s
Asynchronous	RS-232	DB25F	15	DCE	56 kbits/s
Synchronous	RS-422	DB25F	15	Yes	56 kbits/s
Asynchronous	RS-485 RS-530	Terminal block	15	N/A	2.048 Mbits/s
Asynchronous	RS-232 RS-422 RS-485 (2- or 4-wire)	DB25F	15	Yes	600 kbits/s
N/A	T1/E1	Terminal block	>20	N/A	2.048 Mbits/s
N/A	T1/E1	Terminal block	>20	N/A	2.048 Mbits/s
N/A	T3/E3	Coaxial	>15	N/A	2.048 Mbits/s
Asynchronous/ synchronous	RS-232, RS-530, V.35	DB25F	>15	Yes	2.048 Mbits/s

*Data circuit-terminating equipment/data terminal equipment.

The Model TC1540

The TC Communications Model TC1540 fiber-optic modem represents a very versatile copper connection extender. This is due to the ability of the TC1540 to be obtained with an RS-232, RS-422, RS-530, or RS-485 two- or four-wire interfaces for point-to-point communications.

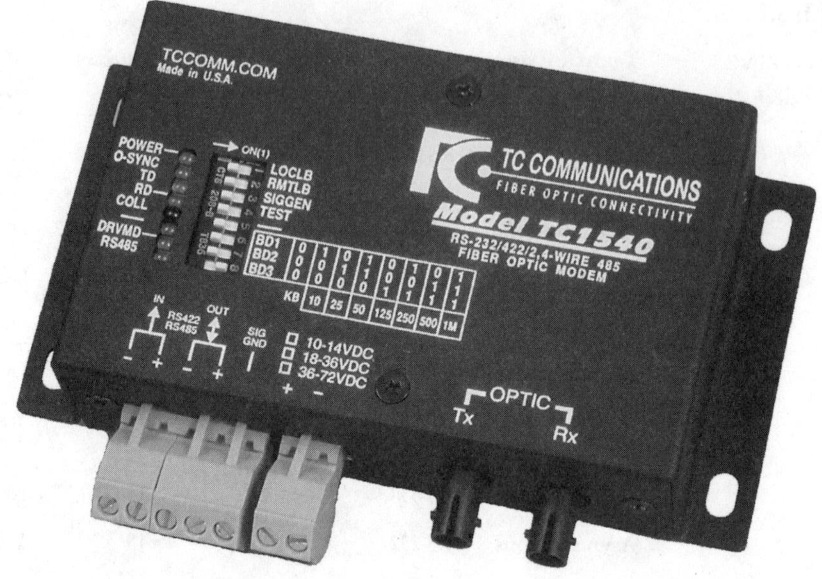

Figure 8.2
The TC Communications Model TC1540 fiber-optic modem. (*Photograph courtesy of TC Communications.*)

If you focus on Figure 8.2, the applicable electrical connector is located below the top edge of the fiber-optic modem. That connector will be a DB25 socket connector for RS-232 or RS-422 (RS-530) interfaces and four terminals for an RS-485 interface capable of supporting two- and four-wire connections.

The TC1540 is powered by DC, with a 9- to 12-V-DC power input connector located below the top right edge of Figure 8.2. Thus, a wall outlet within close proximity of each TC1540 is an important consideration.

If you turn your attention to the left portion of Figure 8.2, you will note a row of eight dual inline program (DIP) switches and eight LED indicators. The DIP switches govern such functions as enabling a local electrical loopback for diagnostic testing, defining the looping of received data to transmit data, generating a simulated 2-Hz transmit pulse signal that is transmitted via the attached fiber cable to a remote unit, setting a transition timer for an RS-485 connection, and enabling primary and secondary optics. The RS-485 transition timer is set by three DIP switches to a predefined value based on the asynchronous data rate to be supported. This setting is required if you are using a two-wire RS-485 interface, as this represents a half-duplex operation in which the transmit and receive operations take turns. Concerning the optical connections supported by the TC1540, if you again refer to Figure 8.2, you will note two pairs of optic connectors at the lower portion of the figure. The pair of connectors on the right side are ST fiber-optic connectors and represent the primary optics that are enabled by one DIP switch. Thus, a pair of TC1540 fiber-optic modems requires two fibers, one for transmission from one modem to the other while the second fiber enables one modem to receive transmission from the other device. The second pair of connectors to the left of the primary connectors represents an optional optic redundancy feature that is included only when a dual-optic TC1540 fiber-optic modem is ordered. Here the second connector pair can be obtained as FC connectors and provides a mechanism for supporting critical applications. The TC1540 supports multimode fiber at wavelengths of 850 or 1310 nm as well as single-mode fiber at wavelengths of 1310 or 1550 nm.

LED INDICATORS The LED indicators visually denote the presence of power to the unit, an alarm indication which results in a flashing signal when the optical signal is lost, as well as the transmission and reception of data on both the electrical interface and optical interface. Thus, the LEDs provide a visual indication of the operational state of the modem. In fact, if you obtain a pair of TC1540 fiber-optic modems with dual-optic connectors and one cable breaks, not only do the modems switch over to the secondary link but, in addition, the alarm LED will flash to indicate the cable break.

TRANSMISSION DISTANCE While the primary use of fiber-optic modems is for use within a building, most such devices permit support for transmission on a university or other campus or even within an industrial park. For example, the TC1540 supports a transmission distance of up to 3 km on multimode fiber at a wavelength of 850 nm and up to 4 km when a wavelength of 1310 nm is used. When single-mode fiber is used at 1310 nm, a transmission distance of 24 km becomes possible. Because the optical loss budget, including connectors and fiber loss, determines whether it is possible to reach a predefined range, it is important to check the connectors as well as consider the total loss of the link. Concerning the latter, Table 8.2 indicates the approximate loss per kilometer that you can use for four different types of optical cable.

TABLE 8.2
Link Loss Values

Fiber	Loss
Multimode, 850 nm	3 dB/km on 62.5/125-μm cable
Multimode, 1310 nm	2 dB/km on 62.5/125-μm cable
Single-mode, 1310 nm	0.5 dB/km on 9/125-μm cable
Single-mode, 1550 nm	0.4 dB/km on 9/125-μm cable

In examining Table 8.2, you need to consider the loss on different fiber in conjunction with transmitter launch power and receiver sensitivity. For example, the TC1540 has a typical launch power of −20 dBm for multimode fiber at both wavelengths (850 and 1310 nm) as well as for insertion into single-mode fiber at a wavelength of 1310 nm. The sensitivity of the TC1540 is −36 dBm for both types of fiber at all three wavelengths.

To facilitate the optical modem selection process, Figure 8.3 includes a list of key device selection features in the left column. The first column to the right of the "Feature" column is labeled "Requirements" and represents a location where you can define your specific requirements for a fiber-optic modem features. The two columns to the right of the "Requirements" column enable you to compare two vendor products against your specific requirements. Of course, you can duplicate this table to compare products from more than two vendors.

Fiber-Optic Multiplexers

A second fiber-optic device that is commonly used for intrabuilding communications is a *fiber-optic multiplexer,* which is the focus of this section. Similar to fiber-optic modems, a wide range of fiber-optic multiplexers have been developed to satisfy a number of different intrabuilding communications requirements.

Rationale

One key difference between a fiber-optic modem and a fiber-optic multiplexer is that the latter is designed to support multiple data sources,

Figure 8.3 Fiber-optic modem selection features form.

Feature	Requirement	Vendor A	Vendor B
Electrical interface			
RS-232	________	________	________
RS-422	________	________	________
RS-485	________	________	________
RS-530	________	________	________
V.35	________	________	________
T1	________	________	________
T3	________	________	________
Other	________	________	________
Fiber Optic Cable support			
Multimode	________	________	________
Wavelength	________	________	________
Single-mode	________	________	________
Wavelength	________	________	________
Fiber optic connectors			
ST	________	________	________
FC	________	________	________
Other	________	________	________
Launch Power			
Single-mode	________	________	________
Multimode	________	________	________
Receiver Sensitivity			
Single-mode	________	________	________
Multimode	________	________	________

which represents the primary function of any multiplexer. Specifically, a multiplexer represents a device that aggregates two or more data sources onto a common communications link.

In a WAN environment, a multiplexer is employed primarily to reduce the number of parallel communications circuits routed between similar locations since subscribers pay a monthly fee based on the length, in mileage, of each circuit and its operating rate. In a building environment, where an organization can install its own communications facilities, it is obvious that they do not pay a repeating fee for each transmission cable installed within the building. However, because the cost of cable over long runs can very easily add up to a significant amount of funds, multiplexers are also used within buildings. In addition, once a conduit within a building is filled to capacity, it is an expensive proposition as well as a time-consuming and potentially disruptive process to install an additional conduit. Because a multiplexer can be used to increase the transmission capacity of existing circuits residing in a conduit, it is often possible to add logical or derived circuits without having to install a new conduit when the existing one is full. In addition, because optical fiber does not transmit electrical signals, it is often possible to string an optical fiber through a building by simply taping the fiber to the exterior of an existing conduit.

Overview

If you are familiar with the operation of traditional time division multiplexers (TDMs), then you will understand that of a fiber-optic multiplexer, which is very similar, except that the latter aggregates input onto an optical fiber instead of having an electrical interface for the high-speed connection. For readers not familiar with the basic operation of a TDM, Figure 8.4 illustrates how it aggregates data from a series of lower-operating-rate interfaces for transmission into applicable time slots on the high-speed interface.

In examining Figure 8.4, note that the letters "A," "B," and so on shown flowing into the TDM located on the left side of the illustration can represent either bytes or bits, depending on the type of multiplexer input aggregation used. The areas without a letter represent the absence of activity at a particular point in time.

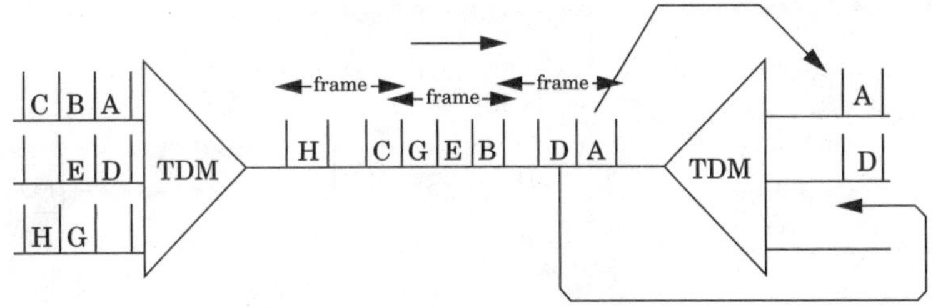

Legend:
TDM Time Division Mutiplexer

Figure 8.4
A time division multiplexer places input data into predefined time slots on a high-speed channel that has a data transfer rate at least equal to the sum of the lower-speed channels supported.

As data are input into the TDM shown on the left side of Figure 8.4, they scan each input, gathering a byte or a bit for formation into a frame for transmission. Because the TDM at the opposite end of the circuit performs demultiplexing by placing all received bytes or bits onto an output channel according to their positions within a frame, spaces or blanks representing the absence of activity are included in a frame. While the TDM in the right portion of Figure 8.4 is shown demultiplexing data, each TDM performs both multiplexing and demultiplexing data by time.

Features

Similar to an optical modem, several features warrant attention when considering the use of a fiber-optic multiplexer. One key feature to consider is the fact that a fiber-optic multiplexer includes a built-in fiber-optic modem. This means that when your organization must acquire the ability to transmit multiple logical channels of data simultaneously over one physical channel, you can consider two equipment configurations. First, you can acquire a pair of traditional TDMs that are restricted to having electrical interfaces. Then you can connect the high-speed interface of each TDM to an optical modem, with each fiber-optic modem connected to a common optical fiber. The top portion of Figure 8.5 illustrates the use of traditional TDMs with fiber-optic modems. In contrast, the lower portion of Figure 8.5 illustrates the use of a pair of fiber-optic modems to accomplish a similar networking

a. Using traditional TDMs and fiber-optic modems.

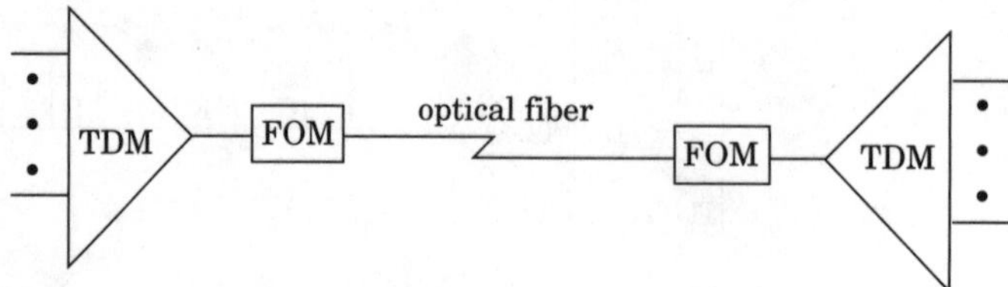

b. Using fiber-optic multiplexers.

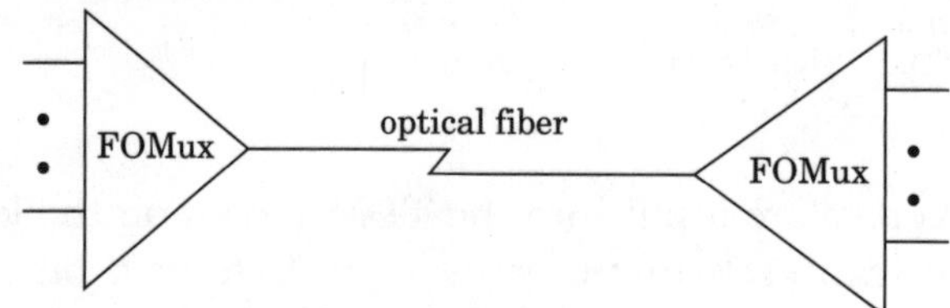

Legend:
TDM time division multiplexer
FOM fiber-optic modem
FOMux fiber-optic multiplexer

Figure 8.5 Two methods are available to transmit multiple channels over a common fiber-optic cable.

capability. The word "similar" is used in this context to reflect the fact that the number of channels supported by traditional TDMs commonly exceeds the number of channels supported by fiber-optic multiplexers. Thus, in certain situations the configuration shown at the top of Figure 8.5 may represent a more flexible network solution to satisfy many organizational requirements.

Actually, there is a third method that will permit multiple channels of data to flow over single optical fiber. That method is *wavelength division multiplexing* (WDM) or its more sophisticated cousin, *dense wavelength division multiplexing* (DWDM). Because both WDM and DWDM are employed primarily on wide area networks, we will not consider this technology for intrabuilding communications; however, this does not mean that it cannot be used.

In addition to considering the number of channels supported by a fiber-optic multiplexer, several additional features warrant consideration. Those additional features include support for asynchronous or synchronous data on each channel, the electrical interface and electrical connector on each channel, and the maximum data transmission rate supported. Because a fiber-optic multiplexer includes a built-in fiber-optic modem, you also need to consider the type of fiber supported, the type of fiber connectors on the fiber-optic modem portion of the multiplexer, and the loss budget.

Representative Multiplexers

To obtain an appreciation for some of the potential functionality of a fiber-optic multiplexer, we will consider two TC Communications products, which represent two variants of the vendor's TC8300 fiber-optic multiplexer.

THE TC8300p The first version of the TC8300 represents a four-channel T1/E1 fiber-optic multiplexer: the TC8300p. Figure 8.6 illustrates the rack-mounted version of the TC8300p. The TC8300p represents a specialized type of fiber-optic multiplexer that can be used to interconnect digital PBXs, channel banks, and similar devices that operate at 1.544 or 2.048 Mbits/s within a building. In addition, the TC8300p can be used to extend the termination point where T1 or E1 channel banks are installed within a building.

The TC8300 supports several types of line coding associated with T1 and E1 copper circuits, such as *alternate mark inversion* (AMI), under which positive pulses are alternated by polarity and the energy of each pulse is concentrated within the middle portion of the pulse area. Two other line coding techniques supported by the TC8300p are the Binary 8 Zero Suppression (B8ZS) and High Density Binary Three (HDB3) coding. Both B8ZS and HDB3 are techniques that look for a group of eight or four consecutive zeros, respectively, and use bipolar violations to

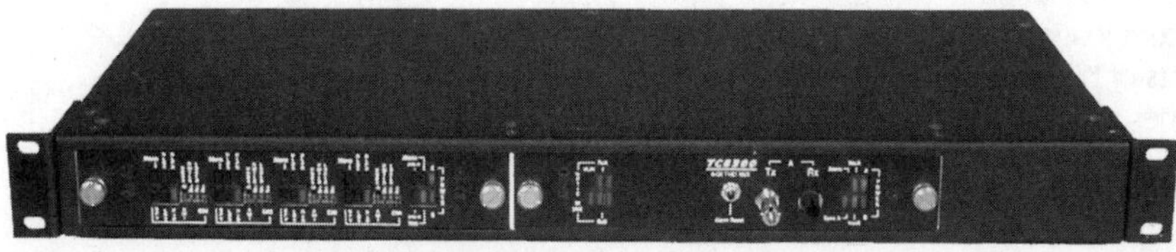

Figure 8.6 The TC8300 is a four-channel T1/E1 fiber-optic multiplexer. (*Photograph courtesy of TC Communications.*)

substitute for a predefined sequence of eight or four binary zeros. The reason for this substitution is the need for T1 and E1 copper circuits to have a minimum "ones" density to ensure that repeaters have a sufficient number of one bits on which to synchronize their operation. B8ZS coding is used on T1 lines, while HDB3 coding is used on E1 lines.

UTILIZATION The TC8300p is well suited for many potential applications. For example, you could use a pair of TC8300p fiber-optic multiplexers to interconnect two or more routers and/or PBXs located in different buildings within a campus or industrial complex where it is possible to install your own fiber. Within a building, the local telephone company will often install a demarcation point where a T1 or E1 line terminates. At that location you could install a T1 or E1 electrical interface fiber-optic multiplexer while locating routers or digital PBXs beyond the relatively short distance normally allowed from the DEMARC point.

Figure 8.7 illustrates an example of how a pair of TC8300p fiber-optic multiplexers could be used to enable routers and a PBX to be located at a considerable distance from the copper DEMARC point where the high-speed lines normally terminate within a building.

In examining Figure 8.7, note that although only three T1 lines are shown as multiplexed, the TC8300p is capable of supporting four T1 or E1 circuits.

Similar to the fiber-optic modem that we described previously, the fiber-optic multiplexer includes various LEDs and DIP switches. The LEDs indicate the loss of a signal, the occurrence of a bipolar violation,

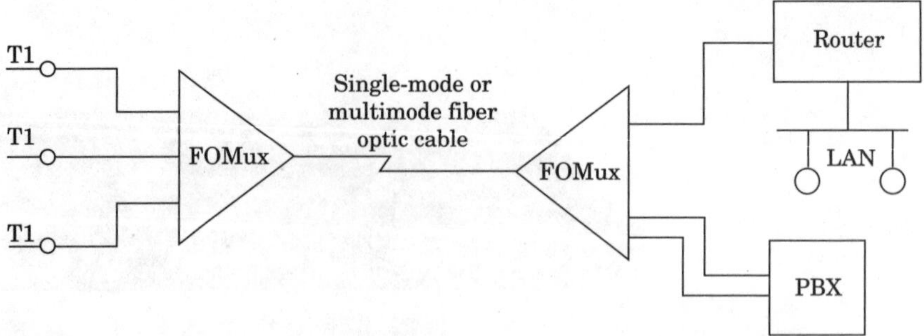

Figure 8.7 Using fiber-optic multiplexers to extend the termination points of high-speed copper circuits.

Legend:
FOMux a TC8300p fiber-optic multiplexer

the status of received data, and other conditions that can be helpful when troubleshooting a problem. In contrast, the DIP switches are used to select a line code for a specific channel or to initial a remote loopback from the local unit or a local loopback for testing the T1 or E1 electrical interface.

The TC8300p is similar to optical modems produced by the vendor in that the fiber-optic multiplexer can be ordered with an optic redundancy option. When an optic redundancy feature is added to the TC8300p, its transmitter sends the same signal via two optical connectors, with the built-in optical modem at the receiving fiber-optic multiplexer determining whether receiver A or B should be used as the valid incoming signal.

TRANSMISSION DISTANCE The transmission distance obtainable between a pair of fiber-optic multiplexers varies in essentially the same way as does the distance between a pair of fiber-optic modems since the former includes the latter. Thus, you must consider the type of fiber-optic cable used as well as the wavelength of the optical signal. For the TC8300p you can obtain a transmission distance of 3 km when a wavelength of 850 nm is used with multimode fiber, while a distance of 4 km becomes achievable when a wavelength of 1310 nm is used. When single-mode fiber is used, a transmission distance of 30 km becomes possible. When a wavelength of 1310 nm is used, a transmission distance of 40 km becomes possible when a lightwave of 1550 nm is used with single-mode fiber.

THE TC8300s A second member of the TC8300 family is the TC8300s. This fiber-optic multiplexer can be obtained with either four or eight RS-232 asynchronous ports and supports a maximum data transmission rate of 56 kbits/s. The TC8300s brings back fond memories for this author, as an equivalent product from another vendor was used in the late 1990s to solve an interconnection problem between two data centers residing in the same building. At that time communications circuits terminated within one data center, with some circuits connected to a switch that could be considered to represent a data PBX (private branch exchange). By connecting asynchronous data circuits to the data PBX, it became possible to allow remote users located around the country to use common long-distance circuits to select their termination points as the data PBX was in turn cabled to several computers in the data center. This situation worked well until it was decided to open a second data center approximately 500 ft down the hall from the first. With no existing conduit available, the use of conventional multiplexers that require the installation of a new conduit would have been disruptive, time-consuming, and expensive. Fortunately, a pair of fiber-optic multiplexers were obtained, which became the solution to the previously described problem.

The RS-232 connectors on one multiplexer were cabled to a group of eight switch ports, with the group of ports configured on the data PBX for routing to the second data center. The RS-232 ports on the second fiber-optic multiplexer were cabled directly to asynchronous ports on a computer located in the second data center, with a fiber-optic cable strung in the ceiling between the two data centers, obviating the necessity to install a conduit. In addition, the use of a pair of fiber-optic multiplexers solved a second problem, one of transmission distance. Because the distance between the two data centers was approximately 500 ft, it was not possible to use conventional multiplexers without inserting at least one repeater between data centers. Because the fiber-optic multiplexer used could easily support a transmission distance of several kilometers, it became possible to simply cable the optical fiber between the two data centers without having to worry about conduits or repeaters. Thus, a fiber-optic multiplexer can represent a versatile communications networking device whose use could solve many communications-related requirements. In fact, one interesting network configuration that deserves mention is the ability of some products to be configured as a ring, with a self-healing capability similar to that of SONET. Thus, before concluding our discussion of fiber-optic modems, let's focus on their self-healing capability.

RING UTILIZATION Depending on the manufacturer, some vendors, including TC Communications, manufacture both fiber-optic modems and fiber-optic multiplexers that can be configured as a self-healing ring. With this ring structure capability it becomes possible to obtain a high degree of redundancy that can provide a capability equivalent to a mini SONET transmission capability within a building or on a college campus or industrial complex.

Figure 8.8 illustrates the configuration of a series of fiber-optic modems and fiber-optic multiplexers in a self-healing ring configuration. Both can be used within a ring if they are produced by the same

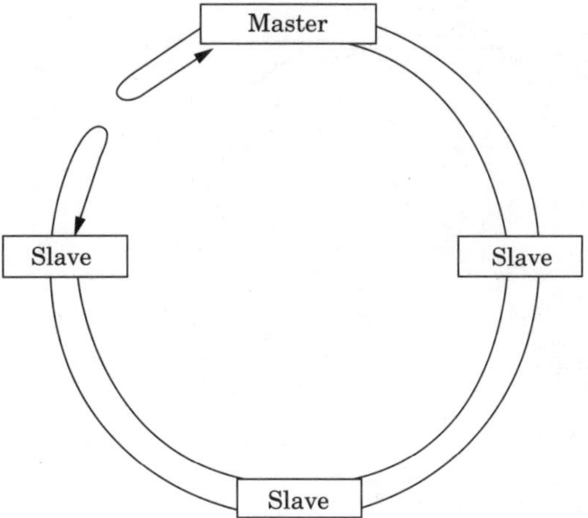

Figure 8.8 Using a ring with self-healing capability.

vendor, have dual-fiber pair connectors, and are manufactured to support a self-healing capability. In this networking environment one device represents a master, while the other devices are slaves. The master periodically polls each slave to determine status. On the detection of a loss of light, the master and slave detecting the loss of signal loop back primary and secondary fiber rings to form a ring.

One innovation that warrants attention is the ability to incorporate multiple master optical modems or optical multiplexers into a self-healing ring. This capability increases the reliability of a ring over the use of a single master as it provides the capability to recover from faults occurring in both rings at two different locations. Thus, the use of multiple masters promotes integrity in the unlikely event of double faults occurring in both the primary and secondary rings. Figure 8.9 illustrates how double breaks in both primary and secondary fibers can be resolved by creating two separate rings.

Optical Mode Converters

In this concluding section we discuss an optical device that you will rarely use; however, when the need arises, you may swear that it is worth its weight in gold. That device is the optical mode converter.

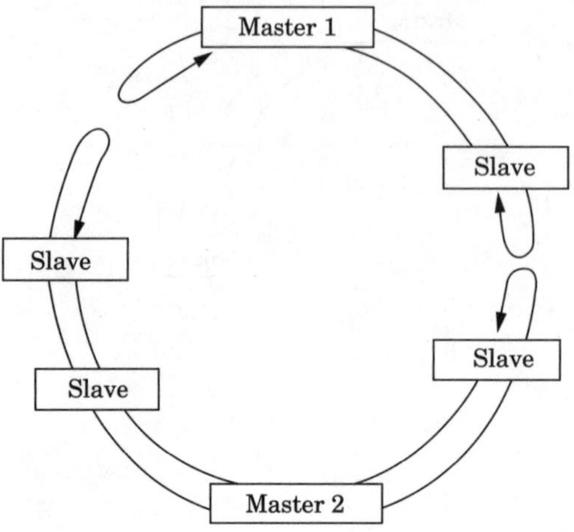

Figure 8.9 Double breaks in both primary and secondary fibers can be resolved by creating two separate rings when there are multiple master devices in a ring.

Overview

One common problem that network managers experience is the need to support a new application while the existing infrastructure consists of optical fiber unsuitable for the application. In fact, this author, while serving as a consultant on a network analysis project for a college, came across this situation. The college had previously installed multimode fiber many years before to support 155-Mbit/s ATM (asynchronous transfer mode) communications from a switch located in one building to all other buildings on the campus. A few years later the college was considering upgrading to Gigabit Ethernet; however, the ability to transmit via multimode using Gigabit Ethernet could not support the distance between the building where the ATM switch was located and several buildings on campus. The solution to this problem was to use multimode where the transmission distance was capable of supporting Gigabit Ethernet. For those locations where the distance precluded the ability to use Gigabit Ethernet or multimode cable, the use of a fiber-optic mode converter provided one solution to the distance problem. This is because some fiber-optic mode converters use LEDs or lasers that have a higher level of optical power than provided by the Gigabit Ethernet standard. Thus, you could use a single-mode/multimode converter to extend the transmission distance. For one product it is possible to obtain a transmission distance of up to 3 km over multimode, which was sufficient for the college.

Operation

Figure 8.10 illustrates the use of a single-mode/multimode fiber-optic converter. Because the converter receives and regenerates light pulses, it also functions as a repeater.

In examining Figure 8.10, note that the single-mode transceiver represents a port on a Gigabit Ethernet switch that would replace the ATM switch used in the campus environment described previously. Here a short single-mode cable is used to connect the switch port to the mode converter, where the latter regenerates the received pulse and transmits it onto the existing multimode fiber-optic cable.

In addition to single-mode/multimode converters, most devices are also capable of performing a reverse conversion. Thus, you can also consider the use of a single-mode/multimode converter as a multimode/single-mode converter.

Features

When selecting a fiber-optic converter, there are several features to consider. Those features include the type of optical cable that the converter interfaces, the wavelengths supported for input and output, the loss budget at a particular wavelength, and the obtainable transmission distance. In addition, you need to consider the optical connector or connectors supported by the converter.

Comparing Converters

In addition to the above-mentioned fiber-optic converter, it should be apparent from reading the chapter covering fiber in the LAN that a

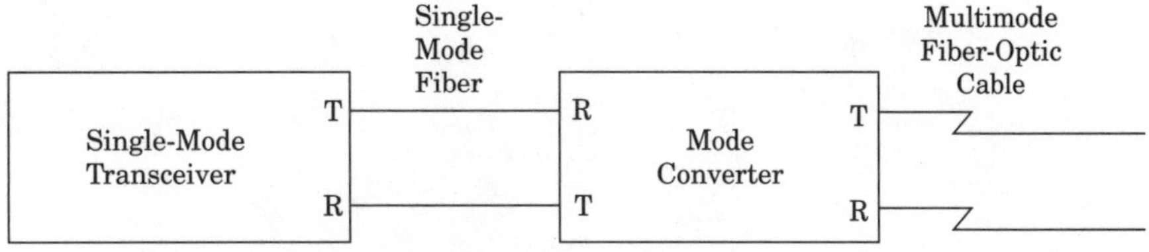

Figure 8.10 Using a single-mode/multimode converter.

second type of converter exists. Thus, in the remainder of this chapter we will compare and contrast the use of the two types of fiber-optic converters.

The converter described in this section represents an optical mode converter. As discussed in this section, this type of converter converts single-mode fiber to multimode and vice versa, and the rationale for its use is its ability to drive the optical signal a sufficient distance beyond the direct use of a particular medium. If we remember our discussion of fiber-optic converters in Chapter 5, we noted that they converted an electrical signal into an optical signal. However, we also noted that their primary role was as a mechanism to extend transmission distance. Thus, the primary difference between the optical mode converter discussed in this section and the media converter discussed in Chapter 5 concerns their interfaces. The mode converter has dual optical interfaces, while the media converter has one electrical interface and one optical interface. However, both types of communications devices are similar in that the primary rationale for their use is to extend transmission distance.

APPENDIX A

FREQUENCY VS. WAVELENGTH

If we were radio listeners we might be familiar with the fact that there are different ways to indicate where to find a certain station on the radio dial. For example, we could say that a station is operating on 4250 kilohertz (kHz), 4.25 megahertz (MHz), or even on 71 meters. And all three ways would be correct!

Radio waves are transmitted as a series of cycles, one after the other. The hertz, named in honor of the German physicist Hertz (abbreviated Hz), is equal to one cycle per second (cps). In fact, if you're as old as the author of this book, you might have used the term *cps* in your high-school physics classes instead of the more modern term Hz.

If you were to visit the fuse box or circuit breakers in your home or apartment you might view a label that indicates the specification for the box or circuit breaker. In addition to telling you information about the amperes and voltage, if you reside in North America, you will also note a reference to the electric power being supplied to your home rated at 60 Hz. Electric power is distributed to your residence as alternating current (ac), meaning it goes through a cycle of changing directions of flow. When we say that electric power is 60 Hz, we mean it changes its direction of flow 60 times per second.

Radio waves cycle more frequently than electricity, resulting in such waves going through far more cycles per second than electric current. Thus, we need to use bigger units to measure them. One such unit is the kilohertz (kHz), which is equal to 1000 cycles per second. Another commonly used unit is the megahertz (MHz), which is equal to 1,000,000 cycles per second, or 1000 kHz. The relationship between these units is shown below:

1,000,000 Hertz = 1000 kilohertz = 1 megahertz

On the wonderful road to optical networking we began to use light, which requires even larger units for expression. This resulted in the use of the term gigahertz (GHz) for representing 1000 MHz. Thus another relation that warrants notice is:

$$1 \text{ GHz} = 1000 \text{ MHz} = 1{,}000{,}000\text{kHz}$$

The term *wavelength* dates from the early days of radio. Back then when life was simple we would gather around the radio in the evening to listen to mystery shows, the top ten songs, and on occasion fireside chats. Back then frequencies were measured in terms of the distance between the peaks of two consecutive cycles of a radio wave instead of the number of cycles per second or the modern-day Hz. What was actually being measured was the wavelength of radio waves, which represents the distance between two troughs or two crests of the radio wave.

During the past century physicists determined the relationship between wavelength and frequency. In doing so they discovered that both the frequency and wavelength were also related to the velocity of propagation of a wave. Physicists determined that the length of a wave multiplied by its frequency gives the velocity at which the wave travels. The equation that relates wavelength, frequency, and velocity is shown below:

$$w \times f = v$$

where w is the wavelength, f is the frequency, and v is the velocity of the wave. For electromagnetic radiation, the velocity of the wave is equal to the speed of light c and the equation becomes:

$$w \times f = c$$

Even though radio waves are like other types of radiation in that they are invisible, there is a measurable distance between the cycles of electromagnetic fields making up such waves. The distance between the peaks of two consecutive cycles of a radio wave is measured in meters. The relationship between a radio signal's frequency in Hz and its wavelength in meters can be found by the following formula:

$$\text{wavelength} = 3 \times 10^8 / f(\text{Hz})$$

In the preceding formula the speed of light is approximately 3×10^8 meters/second. We can also compute wavelength based on frequencies in kHz, MHz, and GHz by varying the numerator and denominator in the prior equation as follows:

$$w\,(\mathrm{m}) = 3 \times 10^5/f\,(\mathrm{kHz})$$

$$w\,(\mathrm{m}) = 300/f\,(\mathrm{MHz})$$

$$w\,(\mathrm{m}) = 0.3/f\,(\mathrm{GHz})$$

In the beginning of this appendix we noted that we could refer to a station on a radio dial in one of three ways. If we use the previously noted relationships between frequency, wavelength, and velocity of light we can now note that a frequency of 4250 kHz would be equivalent to a wavelength of 70.72 meters, which we would round to 71 meters. Thus, the operating terms of 4250 kHz, 4.25 MHz, and 71 meters all refer to the same operating frequency!

As the preceding use of the relationships between frequency, wavelength, and velocity of light indicates, the wavelength of a signal decreases as its frequency increases. Although today frequencies are not commonly given in terms of wavelength in the world of radio communications, in the world of optical communications the two are used interchangeably.

The electromagnetic spectrum is a continuum of wavelengths, from zero to infinity. Physicists name regions of the spectrum based on the type of wavelength in the region. Some of the named regions include: gamma rays, which have the shortest wavelengths (<0.01 nm, where nm stands for nanometer and represents a billionth of a meter) and highest frequency; X-rays that range in wavelength from 0.01—10 nm; ultraviolet radiation, which has wavelengths of 10—310 nm; visible light, which covers the range of wavelengths from 400—700 nm; and infrared, which has wavelengths spanning 710 nm—1 mm. In optical communications we are primarily interested in visible light and infrared light. The wavelength of light is so small that it is conveniently expressed in nanometers (1×10^{-9}) or millimicrons ($\mathrm{m}\mu$), which are equal to one-billionth of a meter or one-thousandth of a micron. The wavelength of violet light varies from about 400 to 450 mm, and of red light from about 620 to 760 mm, or from about 0.000016 to 0.000018 in. for violet, and from 0.000025 to 0.000030 in. for red. In many physics books you will note wavelengths expressed in different terms, with, more than likely, the term used based on the age of the book. At one time it was popular to express wavelengths in terms of cm and in., today it is popular to describe the wavelength of light in terms of nm.

Another term that comes into play when discussing light is the Ångstrom. In the case of light, the wavelength is so short that a specific distance, called the Ångstrom (Å), has been defined as follows:

$$1\text{Å} = 10^{-10} \text{ m or } 10^{-8} \text{ cm}$$

Visible light has a characteristic wavelength in the range of approximately 3900 to 7700 Å. Electromagnetic energy outside this range is no longer visible to the human eye.

Another term sometimes used to express wavelength is the micron (10^{-6} m), which is a thousand times greater than the nm. The term *mu* represents an abbreviation for micrometer and is commonly used by astronomers. In the event you are curious as to why the Greek symbol mu was not used, according to legend it was not available in the typeset used initially by astronomers, resulting in mu being used to form the abbreviation micrometer.

The following table indicates the representative wavelength of visible colors in terms of mu that correspond to our friend, Mr. Roy G. Biv, mentioned earlier in this book.

Color	Representative wavelength (microns)
Violet	0.41
Indigo	0.44—0.45
Blue	0.47
Green	0.52
Yellow	0.58
Orange	0.60
Red	0.65

The following table indicates the relationship between wavelength in nanometers, Ångstroms, and microns. This relationship is for visible light through the infrared range.

Wavelength		
Nanometers	**Microns**	**Ångstroms**
400	0.40	4,000
450	0.45	4,500
500	0.50	5,000
550	0.55	5,500
600	0.60	6,000
650	0.65	6,500
700	0.70	7,000
750	0.75	7,500
800	0.80	8,000
850	0.85	8,500
900	0.90	9,000
950	0.95	9,500
1,000	1.00	10,000
1,050	1.05	10,500
1,100	1.10	11,000
1,150	1.15	11,500
1,200	1.20	12,000
1,250	1.25	12,500
1,300	1.30	13,000
1,350	1.35	13,500
1,400	1.40	14,000
1,450	1.45	14,500
1,500	1.50	15,000
1,550	1.55	15,500
1,600	1.60	16,000
1,650	1.65	16,500

Wavelength		
Nanometers	**Microns**	**Ångstroms**
1,700	1.70	17,000
1,750	1.75	17,500
1,800	1.80	18,000
1,850	1.85	18,500
1,900	1.90	19,000
1,950	1.95	19,500
2,000	2.00	20,000
2,050	2.05	20,500
2,100	2.10	21,000
2,150	2.15	21,500
2,200	2.20	22,000
2,250	2.25	22,500
2,300	2.30	23,000
2,350	2.35	23,500
2,400	2.40	24,000
2,450	2.45	24,500
2,500	2.50	25,000
2,550	2.55	25,500
2,600	2.60	26,000
2,650	2.65	26,500
2,700	2.70	27,000
2,750	2.75	27,500
2,800	2.80	28,000
2,850	2.85	28,500
2,900	2.90	29,000
2,950	2.95	29,500

Wavelength		
Nanometers	**Microns**	**Ångstroms**
3,000	3.00	30,000
3,050	3.05	30,500
3,100	3.10	31,000
3,150	3.15	31,500
3,200	3.20	32,000
3,250	3.25	32,500
3,300	3.30	33,000
3,350	3.35	33,500
3,400	3.40	34,000
3,450	3.45	34,500
3,500	3.50	35,000
3,550	3.55	35,500
3,600	3.60	36,000
3,650	3.65	36,500
3,700	3.70	37,000
3,750	3.75	37,500
3,800	3.80	38,000
3,850	3.85	38,500
3,900	3.90	39,000
3,950	3.95	39,500
4,000	4.00	40,000
4,050	4.05	40,500
4,100	4.10	41,000
4,150	4.15	41,500
4,200	4.20	42,000
4,250	4.25	42,500

Wavelength		
Nanometers	**Microns**	**Ångstroms**
4,300	4.30	43,000
4,350	4.35	43,500
4,400	4.40	44,000
4,450	4.45	44,500
4,500	4.50	45,000
4,550	4.55	45,500
4,600	4.60	46,000
4,650	4.65	46,500
4,700	4.70	47,000
4,750	4.75	47,500
4,800	4.80	48,000
4,850	4.85	48,500
4,900	4.90	49,000
4,950	4.95	49,500
5,000	5.00	50,000
5,050	5.05	50,500
5,100	5.10	51,000
5,150	5.15	51,500
5,200	5.20	52,000
5,250	5.25	52,500
5,300	5.30	53,000
5,350	5.35	53,500
5,400	5.40	54,000
5,450	5.45	54,500
5,500	5.50	55,000
5,550	5.55	55,500

Wavelength		
Nanometers	**Microns**	**Ångstroms**
5,600	5.60	56,000
5,650	5.65	56,500
5,700	5.70	57,000
5,750	5.75	57,500
5,800	5.80	58,000
5,850	5.85	58,500
5,900	5.90	59,000
5,950	5.95	59,500
6,000	6.00	60,000
6,050	6.05	60,500
6,100	6.10	61,000
6,150	6.15	61,500
6,200	6.20	62,000
6,250	6.25	62,500
6,300	6.30	63,000
6,350	6.35	63,500
6,400	6.40	64,000
6,450	6.45	64,500
6,500	6.50	65,000
6,550	6.55	65,500
6,600	6.60	66,000
6,650	6.65	66,500
6,700	6.70	67,000
6,750	6.75	67,500
6,800	6.80	68,000
6,850	6.85	68,500

Wavelength		
Nanometers	**Microns**	**Ångstroms**
6,900	6.90	69,000
6,950	6.95	69,500
7,000	7.00	70,000
7,050	7.05	70,500
7,100	7.10	71,000
7,150	7.15	71,500
7,200	7.20	72,000
7,250	7.25	72,500
7,300	7.30	73,000
7,350	7.35	73,500
7,400	7.40	74,000
7,450	7.45	74,500
7,500	7.50	75,000
7,550	7.55	75,500
7,600	7.60	76,000
7,650	7.65	76,500
7,700	7.70	77,000
7,750	7.75	77,500
7,800	7.80	78,000
7,850	7.85	78,500
7,900	7.90	79,000
7,950	7.95	79,500
8,000	8.00	80,000
8,050	8.05	80,500
8,100	8.10	81,000
8,150	8.15	81,500

Wavelength		
Nanometers	**Microns**	**Ångstroms**
8,200	8.20	82,000
8,250	8.25	82,500
8,300	8.30	83,000
8,350	8.35	83,500
8,400	8.40	84,000
8,450	8.45	84,500
8,500	8.50	85,000
8,550	8.55	85,500
8,600	8.60	86,000
8,650	8.65	86,500
8,700	8.70	87,000
8,750	8.75	87,500
8,800	8.80	88,000
8,850	8.85	88,500
8,900	8.90	89,000
8,950	8.95	89,500
9,000	9.00	90,000
9,050	9.05	90,500
9,100	9.10	91,000
9,150	9.15	91,500
9,200	9.20	92,000
9,250	9.25	92,500
9,300	9.30	93,000
9,350	9.35	93,500
9,400	9.40	94,000
9,450	9.45	94,500

Wavelength		
Nanometers	**Microns**	**Ångstroms**
9,500	9.50	95,000
9,550	9.55	95,500
9,600	9.60	96,000
9,650	9.65	96,500
9,700	9.70	97,000
9,750	9.75	97,500
9,800	9.80	98,000
9,850	9.85	98,500
9,900	9.90	99,000
9,950	9.95	99,500
10,000	10.00	100,000
10,050	10.05	100,500
10,100	10.10	101,000
10,150	10.15	101,500
10,200	10.20	102,000
10,250	10.25	102,500
10,300	10.30	103,000
10,350	10.35	103,500
10,400	10.40	104,000
10,450	10.45	104,500
10,500	10.50	105,000
10,550	10.55	105,500
10,600	10.60	106,000
10,650	10.65	106,500
10,700	10.70	107,000
10,750	10.75	107,500

Wavelength		
Nanometers	**Microns**	**Ångstroms**
10,800	10.80	108,000
10,850	10.85	108,500
10,900	10.90	109,000
10,950	10.95	109,500
11,000	11.00	110,000
11,050	11.05	110,500
11,100	11.10	111,000
11,150	11.15	111,500
11,200	11.20	112,000
11,250	11.25	112,500
11,300	11.30	113,000
11,350	11.35	113,500
11,400	11.40	114,000
11,450	11.45	114,500
11,500	11.50	115,000
11,550	11.55	115,500
11,600	11.60	116,000
11,650	11.65	116,500
11,700	11.70	117,000
11,750	11.75	117,500
11,800	11.80	118,000
11,850	11.85	118,500
11,900	11.90	119,000
11,950	11.95	119,500
12,000	12.00	120,000
12,050	12.05	120,500

Wavelength		
Nanometers	**Microns**	**Ångstroms**
12,100	12.10	121,000
12,150	12.15	121,500
12,200	12.20	122,000
12,250	12.25	122,500
12,300	12.30	123,000
12,350	12.35	123,500
12,400	12.40	124,000
12,450	12.45	124,500
12,500	12.50	125,000
12,550	12.55	125,500
12,600	12.60	126,000
12,650	12.65	126,500
12,700	12.70	127,000
12,750	12.75	127,500
12,800	12.80	128,000
12,850	12.85	128,500
12,900	12.90	129,000
12,950	12.95	129,500
13,000	13.00	130,000
13,050	13.05	130,500
13,100	13.10	131,000
13,150	13.15	131,500
13,200	13.20	132,000
13,250	13.25	132,500
13,300	13.30	133,000
13,350	13.35	133,500

Wavelength		
Nanometers	**Microns**	**Ångstroms**
13,400	13.40	134,000
13,450	13.45	134,500
13,500	13.50	135,000
13,550	13.55	135,500
13,600	13.60	136,000
13,650	13.65	136,500
13,700	13.70	137,000
13,750	13.75	137,500
13,800	13.80	138,000
13,850	13.85	138,500
13,900	13.90	139,000
13,950	13.95	139,500
14,000	14.00	140,000
14,050	14.05	140,500
14,100	14.10	141,000
14,150	14.15	141,500
14,200	14.20	142,000
14,250	14.25	142,500
14,300	14.30	143,000
14,350	14.35	143,500
14,400	14.40	144,000
14,450	14.45	144,500
14,500	14.50	145,000
14,550	14.55	145,500
14,600	14.60	146,000
14,650	14.65	146,500

Wavelength		
Nanometers	**Microns**	**Ångstroms**
14,700	14.70	147,000
14,750	14.75	147,500
14,800	14.80	148,000
14,850	14.85	148,500
14,900	14.90	149,000
14,950	14.95	149,500
15,000	15.00	150,000

GLOSSARY

atom a particle consisting of an inner nucleus surrounded by electrons that circle the nucleus.

attenuation a reduction of optical power as photons flow down fiber.

attenuation coefficient a loss of optical power per a given length of fiber, commonly expressed in dB/km.

backbone cabling cabling between telecommunications rooms.

Baud a rate of signal change commonly expressed in terms of Hz.

bel a power ratio named in honor of Alexander Graham Bell. The bel equals 10 $\log_{10}(P_O/P_I)$ where P_I is power output while PI is the input or transmitted power.

broadband WDM a small section of a fiber modified to create changes in the index of refraction that reflects certain wavelengths.

Bragg gratings a small section of fiber modified to create changes in the index of refraction that reflects certain wavelengths.

carrier an oscillating wave which is modulated to convey information.

centralized cabling cabling from a work area routed to a central location.

chromatic dispersion a change in the speed of optical pulses as their wavelength changes, resulting in a broadening of pulses.

cladding the area between the core and outer area of a fiber that coats the fiber.

Class I repeater a Fast Ethernet repeater that is capable of connecting segments using different coding.

Class II repeater a Fast Ethernet repeater that is only capable of connecting segments that use the same signaling method.

coherent light light in which emitted photons travel in the same wave pattern.

collimation the process of converting divergent light into a beam of parallel light.

connector a mechanical device physically connected to the end of a fiber-optic cable.

critical angle the angle at which light is totally refracted back into the core of a fiber.

cross-phase modulation the broadening of light pulses in a dense wavelength division multiplexing environment.

coupling method the method by which connectors mate. The coupling method includes threaded, bayonet, and push-pull.

dark fiber fiber strands not yet used or illuminated.

decibel a more precise power measurement than bel. A decibel equals one-tenth of a bell.

demarcation the point where a circuit enters a building.

dibit coding the process of encoding two bits in one signal change.

direct ray the flow of light through a fiber on a direct route from end to end.

dispersion the smearing or broadening of an optical signal.

electromagnetic wave a continuum of oscillating electric and magnetic fields moving in a straight line at a constant velocity.

electrovolt energy gained by an electron that passes across a positive voltage of one volt.

fiber mile the length of fiber conduit times the number of strands in a conduit.

fiber-optic modem the use of an LED or laser and photodetector in a common housing to provide electrical-to-optical and optical-to-electrical conversion.

forward bias the application of a positive voltage across an appropriately doped p-n semiconductor.

frequency the number of periodic oscillations or waves that occur per unit time.

frequency division multiplexing an analog technology in which the bandwidth of a transmission medium is subdivided by frequency.

frequency spectrum another name for the electromagnetic radiation spectrum.

Fresnel reflection loss the reflection of light back towards the transmitter due to an air gap between connectors.

gain the amount by which the intensity of light is increased in a laser diode by amplification.

gamma rays radiation with the highest frequency and shortest wavelength.

glass fiber a cable with a glass core and glass cladding.

graded-index fiber a fiber whose refractive index gradually diminishes from the center of the core outward towards the cladding.

heterojunction a crystal placed between two dissimilar semiconductors that confines light to the region of a p-n junction in a semiconductor.

horizontal cabling cabling between a telecommunications outlet in a work area and the horizontal cross-connect.

incoherent light that radiates in all directions.

index of refraction the ratio of the velocity of light in a vacuum to that in a material.

infrared waves above the microwave region but below visible light.

insertion loss a loss of optical power at a connector measured in dB.

jacket a coating applied over the cladding of a fiber that represents the outer layer of the cable.

laser a device that generates coherent light.

light-emitting diode a semiconductor device that emits incoherent optical radiation in the form of electromagnetic waves when biased in the forward direction.

lighted fiber fiber illuminated and in use.

lit fiber a fiber strand in use.

material dispersion the non-linear relationship between frequency and the index of refraction in the core of a fiber that broadens an optical pulse.

media interface connector a connector used to connect multimode fiber to an FDDI station.

microwaves waves shorter than radio waves but longer than infrared.

modal dispersion the spreading of light pulses.

mode field diameter a measure of the intensity of light traveling within a fiber.

modified chemical vapor deposit the deposit of chemical vapor or gas during the fabrication of an optical fiber.

multimode fiber fiber with a relatively large core, which allows multiple rays of light to simultaneously propagate through the fiber.

Newton's prism a prism used by Isaac Newton to analyze the colors of light.

non-return-to-zero a transmission scheme by which 0 is coded as a 0 signal level and 1 as a plus pulse

numerical aperture the square root of the difference between the square of the core of a fiber and the square of its cladding.

Nyquist relationship the relationship between bandwidth (W) and the signaling rate (B) such that B = 2W.

optical mode converter a converter that converts single-mode fiber to optical mode fiber and vice versa.

optical window a wavelength or group of wavelengths.

peering point location where two or more Internet Service Providers interconnect.

period the time required for a signal to be transmitted over a distance of one wavelength.

photodetector a device that detects light and generates an electric current.

photoelectric effect the transfer of light energy to an electron when a metal placed in a vacuum has light focused.

photon a Greek word meaning light.

photophone a device invented by Alexander Graham Bell that used a membrane to modulate an optical signal.

plastic-clad silica an optical cable manufactured using a glass core and a plastic coating.

power budget the difference in dB between transmitted optical power and receiver sensitivity.

quantum efficiency the ratio of the number of emitted photons to the number of electrons that cross the p-n junction of a semiconductor.

radio waves waves that have the longest wavelength and lowest frequencies of all waves within the frequency spectrum.

Raman effect amplification that occurs when a large continuous-wave laser is co-launched at a lower wavelength than the signal to be amplified.

Rayleigh scattering a phenomenon recognized by Lord Rayleigh that notes that short wavelengths scatter more strongly than longer wavelengths.

scattering the spreading of optical pulses as photons collide with impurities in a fiber.

section under SONET a section is a transmission path between two repeaters.

self-phase modulation the broadening of light pulses as they flow down a river.

Shannon's Law a formula that defines the maximum bit rate on a channel in terms of bandwidth and the signal-to-noise ratio.

shell the area of an atom where electrons circulate.

signal-to-noise ratio the ratio of signal power divided by noise power on a transmission medium.

single mode a flow of light that lacks refraction, resulting in pulses traveling the fiber in one mode.

step-index fiber a fiber which has an abrupt change in the index of refraction going from the core to the cladding.

tempest electronic emissions generated by equipment.

total internal reflection the refraction of a beam of light so that virtually 100 percent remains in a medium.

traverse burning the rotation of a large glass tube over a heat source while passing a series of gases through the tube.

tunable laser a laser capable of transmitting multiple bands of light through a single strand of optical fiber.

ultraviolet radiation in the electromagnetic spectrum above visible light.

valance band the outer shell of an atom.

vertical-cavity laser a laser formed by the vertical stacking of mirrors.

visible light wavelengths between 400 and 700 nm.

wavelength the period of an oscillating signal.

wavelength division multiplexing the process of transmitting multiple optical signals at different wavelengths over a common optical fiber.

white light a term used to refer to visible light.

white noise background noise generated by the movement of electrons.

x-rays high energy waves that reside above the ultraviolet band in the frequency spectrum.

4B/5B coding a coding technique by which 4 bits are coded using a 5-bit pattern.

ABBREVIATIONS

A	attenuation
ADSL	Asymmetrical Digital Subscriber Line
AM	Amplitude Modulation
ANSI	American National Standards Institute
ATM	Asynchronous Transfer Mode
AUI	Attachment Unit Interface
B	bel
BD	beam diameter
B8ZS	binary eight zero suppression
CIA	Central Intelligence Agency
CIE	Commission Internationale l'Eclairage
CIR	Committed Information Rate
CLEC	Competitive Local Exchange Carrier
CPS	cycles per second
CSMA/CD	carrier sense multiple access with collision detection
DA	full-divergence angle
DACCS	digital access and cross connect system
dB	decibel
dBm	decibel-milliwatt
DEMARC	demarcation
DIP	dual inline program
DOCSIS	Data Over Cable System Interface Standard
DSU	Digital Service Unit
DWDM	Dense Wavelength Division Multiplexing
ECSA	Exchange Carriers Standards Association
EDFA	erbium-doped fiber amplifier
EMI	electromagnetic interface
eV	electronvolt

FCC	Federal Communications Commission
FDDI	fiber distribution data interface
FDI	fiber distribution interface
FM	Frequency Modulation
femto	quadrillionth
FN	fiber node
FOIRL	fiber-optic inter-repeater line
FSK	frequency-shift keying
FTTC	fiber to the curb
FTTH	fiber to the home
FWM	four wave mixing
giga	billion
HBO	Home Box Office
HFC	hybrid fiber-coaxial
Hz	Hertz
ISA	Industry Standard Architecture
IP	Internet protocol
JPEG	Joint Photographics Expert Group
kilo	thousand
LAN	Local Area Network
LASER	light amplification by stimulated emission of radiation
LED	Light Emitting Diode
LLC	logical link control
LWIR	longware infrared region
MAC	media access control
MAN	metropolitan area network
MCVD	modified chemical vapor deposition
mega	million
MEMS	microelectromechanical switch
MFD	mode field diameter
MIC	Media Interface Connector

micro	millionth
milli	thousandth
mW	milliwatt
MWIR	midwave infrared region
NA	numerical aperture
NDSF	non-dispersion-shifted fiber
NRZ	non-return-to-zero
NRZI	non-return-to-zero-inversion
NZDSF	non-zero-dispersion-shifted fiber
OC	optical carrier
OSI	Open System Interconnection
OVD	outside vapor deposition
PBX	private branch exchange
PCI	Peripheral Component Interconnect
PCM	pulse-code modulation
PCS	plastic-clad silica
Pico	trillionth
PM	Phase Modulation
PSTN	Public Switched Telephone Network
S/N	Signal-to-noise ratio
SD	spot diameter
SDH	Synchronous Digital Hierarchy
SMA	sub-multi-assembly
SMT	station management
SONET	Synchronous Optical Network
SPE	synchronous payload envelope
ST	straight tip
STS-1	Synchronous Transport Signal level 1
SWIR	short-wave infrared region
TDM	time division multiplexer
tera	trillion

TIR	total internal reflections
TTP	timed token protocol
TVX	valid transmission timer
VCSEL	vertical-cavity surface-emitting laser
WAN	wide area network
WDM	Wavelength Division Multiplexing

INDEX

A

B

C

D

E

F

G

About the Author

Gilbert Held is an internationally recognized author and lecturer who specializes in the application of computer and communications technology. The author of over 40 books covering personal computers, data communications, and business trade topics, he is the only person to twice win the competitive Karp Interface Award. A winner, as well, of the American Publishers Institute and numerous other industry awards, Mr. Held was selected by *Federal Computer Week* as one of the top 100 persons in government, industry, and academia who has made a difference in the acquisition and use of computer systems. Among his other books are *Voice Over Data Networks; Cisco Security Architecture; Cisco Router Performance Field Guide*; and *Voice/Data Internetworking*, all from McGraw-Hill.